Jack Steinberger
Learning About Particles – 50 Privileged Years

Jack Steinberger

Learning About Particles – 50 Privileged Years

With 117 Figures

Dr. Jack Steinberger
CERN LEP
1211 Geneve
Switzerland

Library of Congress Control Number: 2005105853

ISBN 3-540-21329-5 Springer Berlin Heidelberg New York

Springer is a part of Springer Science+Business Media
springeronline.com

Printed in Germany

Typesetting: Author's data, final processing by LE-TEX Jelonek, Schmidt & Vöckler GbR, Leipzig
Cover design: Erich Kirchner, Heidelberg

Printed on acid-free paper 57/3141/YL 5 4 3 2 1 0

Prelude

One of the great cultural achievements of the twentieth century was the progress in our understanding of particles and their interactions. Everything in the universe is determined by these particles and their interactions with each other, whether it is the formation of galaxies or the behavior of our cat. The first particle discovered was the electron, by J. J. Thompson in 1897. The photon was recognized as a particle by Einstein in 1904, the proton was discovered by Rutherford in 1914, and the neutron by Chadwick in 1932. The same year saw the discovery, by Anderson, in a cloud chamber exposed to cosmic rays, of the positron, the antiparticle of the electron (one of the fundamental symmetries in nature is that of particles and antiparticles, which have the same mass and angular momentum but the opposite charge). In 1933, to conserve energy and angular momentum in the β decay of radioactive nuclei, Wolfgang Pauli proposed the neutrino (now the electron neutrino), followed within a year by Enrico Fermi with a very successful theory of the new "weak" interaction, the interaction that underlies radioactivity. The first unstable elementary particle, the mesotron (now called muon) with a mean life of two millionth of a second, was discovered in 1937, again in cloud chambers exposed to cosmic rays, by Anderson and Neddermeyer and by Street and Livingston. These were the particles that existed when I entered particle physics in 1947 as a graduate student at the University of Chicago. Now it is known that the proton and neutron are not elementary particles, but are composed of quarks. In addition to these very important beginnings of particle physics, the first half of the twentieth century was marked by two most fundamental vital breakthroughs in physics: the discovery of the theory of general relativity in 1914 and of quantum mechanics and quantum field theory in the twenties and thirties, the latter was the basis for any understanding of the dynamics of particles.

The second half of the century, with the advent of particle accelerators of rapidly increasing energies and intensities, saw beautiful progress in our understanding of the particles and their relationship to one another. We now know that the elementary fermions (fermions are particles with angular momentum $1/2\times$ the Planck constant divided by 2π) consist of three "families" of four members each, each family with increasingly higher masses. The first family consists of the electron neutrino, the electron, the *up* and the *down* quarks; the second family of the muon neutrino, the muon, the *charmed* and the *strange* quark; and the third family of the tau neutrino, the tau, the *top* and the *bottom* quark. In going from one family to another, the masses change rapidly; for instance the tau, third family counterpart of the electron, has more than 3,000 times the electron mass, but the electric charges and the interactions are the same. We know that there are three interactions, each propagated by vector particles (vector particles are particles of angular momentum equal to 1 x the Plank constant divided by 2π): the *electromagnetic* interaction propagated by the mass-less photons, the *weak* interaction propagated by the very massive Z^0, W^+, and W^- mesons, about 200,000 times as massive as the electron, and the *strong* interaction by 8 mass-less, "colored" gluons. The last of these particles to be seen, the tau neutrino, was found only just at the turn of the century, in the year 2000, although no one could doubt its existence since the seventies. The late sixties, early seventies saw the emergence of a beautiful, unified theory of the electromagnetic and weak interactions, the "electro-weak" theory, which has been verified in succeeding years with impressive precision. Nineteen seventy-three saw the consolidation of a theory of the strong interaction, named "quantum-chromo-dynamics," very similar in its mathematical structure to the electro-weak theory, and which has also been verified in succeeding years, although, for technical reasons, not as precisely. Together, the electro-weak and quantum-chromo-dynamics theories form what we call our "standard model." Following the discovery of the first hadronic meson in Bristol in 1947, the pi meson, now known to be composed of a quark and an anti-quark, several dozens of such hadronic particles, both mesons consisting of a quark and an anti-quark, and barions, such as the neutron and proton, composed of three quarks, as well as their antiparticles, the antibarions, such as the antiproton, composed of three anti-quarks, have been discovered and intensively studied, with consequent improvement in our understanding of the strong interaction. All presently observed phenomena can be understood in terms of the known particles and their interactions within the theoretical frame of the standard model.[1]

[1] This is almost true, but not quite, in more than one way. To understand gravitational forces in the universe, it is necessary to postulate matter of an unknown type, the so-called "dark matter," which dominates the total matter content of the present universe. However, extensions of the standard model have been proposed over the last several decades, motivated by difficulties of the standard model at much higher energies. These predict particles which may very well account for the cosmic dark matter, and may even be discovered at the next, very much higher energy accelerator now under construction at CERN in Geneva. Another cosmological need is that of a particle field necessary to produce the inflationary beginnings of the Big Bang. A third deficiency is the missing Higgs particle.

It has been my privilege to be immersed in this beautiful progress and to contribute to it during a long professional career of fifty-odd years. It began with my graduate studies at the University of Chicago, immediately following World War II, where I had the immense luck to find Fermi, one of the outstanding founders of this progress, as my master. Again by chance, I found myself in 1949 at Berkeley, at that time the only place in the world with accelerators that could study the physics of the newly discovered mesons. And so on, in the years that followed, I was always at or near laboratories which permitted this research at the most advanced levels, and with very fine colleagues. This has been my professional life, and now that I am more than eighty, I have the need to write this down. The story is not likely to be interesting to many. It is a small part of the history of the evolution of our knowledge of particle physics, but physics goes on, and today's problems are not those of yesterday. I hope that nevertheless there might be a few with the leisure and interest to read a bit of what happened then. Perhaps, to the extent that my story is also connected with some rather grotesque events of those years, such as Nazi Germany and McCarthy U.S.A., this may add some interest.

This is a spin zero (so-called scalar) particle necessary for the consistency of the electro-weak theory.

Contents

1

Origins and Education

1.1 Franconian Beginnings

My parents were of the German-Jewish middle class. Father was born in 1874, one of eight children, in the small Lower Franconian peasant village of Schonungen. His father, making his living as a cattle dealer, was of the lower middle class. Father went to public school, then to Jewish seminary, and at 18 became a cantor and Hebrew teacher in nearby Bad Kissingen, a job which he held until his emigration to America in 1937, 45 years later. Bad Kissingen was, and still is, a popular health resort with mineral springs that are supposed to cure many ills and had a rather large Jewish community of some 300 souls. The job was half paid by the congregation, the cantor part, the teacher half was paid by the state, since religious education in primary school was mandatory. It was interesting that Father continued to receive his state salary after the Nazis took power in 1933 until his emigration in 1937, despite the virulent anti-Semitism.

Father was easygoing, content in his work and life, without ambition despite a great intelligence and clarity of understanding of what went on around him, generous to all, and much appreciated by those who knew him. During the war of 1914–18 he served in the Bayrischer Landsturm, mostly in occupied France. One year later, at the age of 45, he married, perhaps with the help of a matchmaker, as was not unusual then in the Jewish community. Mother, 18 years his junior, came from a more well-to-do family. Her father was a hop merchant in Nuremberg and owned a house near the center of this beautiful, historic town. She had been able to attend university, not common for women at that time; her English and French studies included stays in London and Paris. Also, Mother was uncommonly intelligent, and by nature kind, but with a critical mind that drove her sometimes to say things that were not appreciated. Unfortunately, my own character turned out more in the image of mother than father. My parents, as best I know, got on in excellent harmony; I never heard them quarrel.

The Bad Kissingen Jewish community had both well-to-do and simple members. Some owned hotels that catered to the Jewish tourists, others were shopkeepers, a few were doctors, one made mattresses, one was a jeweller, one a banker, one a junk collector, some were cultured, others less. Although Jews, in those years before the seizure of power by Hitler, were in many ways integrated and respected in the larger community, they were dependent on each other for their social lives. Our family's income and lifestyle were modest, but did include a maid, as was common in those days. Father played piano and earned a bit of extra money giving piano lessons; for some time he led the town chorus. One of the things I have not understood is why he did not succeed in teaching piano to any of his three sons. Mother gave private lessons in English and French to occasional tourists as well as to her children.

We lived in the Jewish community house on the same grounds as the large synagogue. The house consisted of two wings; on one side was the Beetsaal, that is the room for the daily religious services attended only by a small number, and below, the schoolroom in which my father taught. The other wing contained apartments for three of the families that served the community, in addition to ours: the ritual slaughterer and the non-Jewish caretaker of the synagogue and grounds. In the basement there was the ritual bath for women as well as the slaughter room for fowl. There was also a rabbi, but he lived elsewhere.

The religious services, which my father led and sang, were in the German-Jewish orthodox tradition, but the members of the community were of all shades of compliance with the many rules; some, as ours, observed the dietary laws carefully, as well as the restrictions on the activities permitted on the Sabbath. For instance, since it was not permitted to light a gas flame on the Sabbath, we had a substantial cooking device in the kitchen, a kind of oven, heated with carbon, which stayed hot for the 24 hours needed. Other families, particularly some of the more wealthy, kept few of the rules, came rarely to the synagogue, but were members of the community just the same. As part of the expected lifestyle, father regularly, perhaps for an hour or so several times per week, would study the Talmud. But there was a definite cleavage in style and attitudes even within the orthodox community, between the more traditional rabbi and his followers and my more liberal father and his group.

My elder brother Herbert and I were separated by one year, Rudolf came three years later. As it happened, the age rules for admission to primary school were changed at the time of my admission. As a result, my elder brother and I found ourselves in the same grade and classroom. In those years, after four years of Volksschule, it was necessary to choose between continuing on a road with very limited education, unfortunately the more common road, or switching to what in Kissingen then was the Realschule and is now the Gymnasium. The competition for best grades in our classes was limited to the two brothers, sometimes it was Herbert, sometimes it was Hans, then my name. Otherwise, we were different: Herbert enjoyed serious literature, I did little reading, but made things with my hands, modelled in clay and painted.

The late twenties and early thirties witnessed intense and often violent political struggles between the upcoming National Socialists and the Left. The anti-Semitism of the Nazis was far from hidden, rather it was brandished in the most virulent way as part of the campaign to win adherents. I remember, from those years before Hitler had seized power, nighttime torchlight parades of the uniformed, brown-shirted stormtroop SA militia, singing the Horst Wessel song, with the refrain "wenn's Judenblut vom Messer fliesst, dann geht's nochmal so gut" (when the blood of the Jews is flowing from our knives, then things are really going well). The daggers were on their belts. I wish that I had a better memory of the reactions that these scenes, as well as the many political posters with the most awful caricatures of vile Jews, evoked in the mind of a 10- to 12-year-old child. But in school we were still treated very correctly, partly because of the care of the principal, Herr Dr. Hoffmann, who deliberately and courageously protected us; this cost him his job some years later. Personally, I remember no anti-Semitic acts against me by classmates. It would be interesting for me to know how my father reacted to this anti-Semitism. As with most of the German Jews, he had seen himself as Deutscher Staatsbürger Jüdischen Glaubens, a loyal German, glad to have served his country in the war. I myself, as a young boy, had been brought up according to the doctrine that the Vaterland, as a matter of course, can count on us to sacrifice our lives if the occasion arises, and that we Germans are better than others, that we work harder, are more honest and more courageous. This unchallenged nationalistic dogma, with which the German was indoctrinated at that time, is part of the background that may allow one to understand the enthusiastic way in which most Germans followed Hitler to anti-Semitism, to war and, in millions, to their death. Hitler did not invent this nationalism, it was part of the culture; nor did he invent the anti-Semitism, it was completely prevalent in the pre-Hitler right-wing, ultra-nationalist movements. Even the swastika was already being used by these movements. But the pre-Hitler ultra-nationalism was based on the old nobility, landowning, and bourgeois classes; the great invention of the Nazis was to solicit the additional support of the working class.

Following the Nazi power takeover, the conditions of the Jewish community rapidly deteriorated. Within months the population was exhorted to boycott stores and businesses owned by Jews, and Jews were eliminated from public positions. In 1935, the Nuremberg laws were enacted. Jews were defined in terms of the number of "Jewish" grandparents, and were disenfranchised. Sexual relations between "Jews" and "Aryans" were forbidden. As early as April of 1933, a law was passed against the overcrowding of German schools, which limited the enrollment of Jewish children in public schools, and in 1938, Jewish attendance in public schools was forbidden. By the spring of 1934, the situation had deteriorated to the point where my parents decided to accept, for their two older boys, now 13 and 14, an offer by the American Jewish charities to take charge of some 300 teenage German Jewish children and to try to find foster homes for them in the U.S. It was clearly not an easy decision for parents to separate the family; our parents were among the first to sense the gravity of what was happening. The transport was arranged for December. Our belongings were packed in two large, flimsy trunks bought for the purpose, and in addition we could take our bicycles. The whole family came on the trip as far as Bremen. This included a stopover of some days in the big city of Berlin, where, to escape terrorist threats, my mother's brother, a doctor, had fled from Nuremberg some months before and opened a practice. We visited some of the museums.

Late in 1988, following the announcement of that year's Nobel prizes, I was able to revisit the scenes of these early years, which I had fled in 1934, on the invitation of the town authorities and teachers at the Gymnasium. The synagogue was gone, a victim of the Kristallnacht pogrom of 1938; it has been replaced by an ugly municipal office building. The house of the former Jewish community, next door to where the synagogue had stood, the house in which I was born and grew up, was still there, maintained by the town in memory of the former Jewish community, although no Jews were living in Bad Kissingen at that time. The Beetsaal, which had been used for the small weekday services, had been refurbished with the help of the reborn Jewish community of the nearby district capital Würztburg, and was used to offer religious services for Jewish guests during the summer. On the front of the house the town has placed a bronze plaque in memory of 69 concentration-camp victims. Many of the names on this plaque are memories of my childhood.

The year of my return was the fiftieth anniversary of the Kristallnacht. The occasion had been used by two of the Gymnasium teachers to lead some of their students in a historical study of the former Jewish community of their town, and the material that they had found had been put on public display in the beautiful old Rathaus. I could recognize pictures from my youth, including one of my parents. Maintained and augmented by the dedication and devotion of the teachers and students, it is now on permanent display in the old Jewish community house. The story of the Jews of Bad Kissingen is typical of that of many similar such communities in Germany, and the book which has been published by Hans-Jürgen Beck and Rudolf Walter on Jewish life in Bad Kissingen [1] should be a useful reference for anyone interested in this culture, which produced such men as Marx, Einstein, Freud, and Mahler and which is now extinct. Of course, there is nothing special about the extinction of cultures, this is their normal fate.

Since then, I have returned to Bad Kissingen several times, once to give a lecture on supernovae in the Gymnasium. I am very conscious and appreciative of the rediscovered connection with the environment of my early years and of the contacts and friendships that have developed with several of its present members, who include the above teachers but also the then Oberbürgermeister and the Oberstudiendirektor of the Gymnasium, as well as others, all deeply committed to the importance of not forgetting the atrocities of the Nazi years

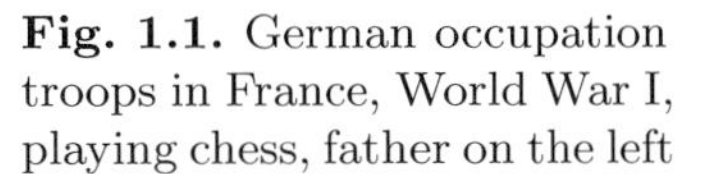

Fig. 1.1. German occupation troops in France, World War I, playing chess, father on the left

and of rebuilding a German society as free from racial prejudice as possible (Figs. 1.1–1.4).

1.2 The New World

The mid-winter Atlantic crossing from Bremen to New York, on the ten-thousand-ton U.S. Line *SS United States*, included some of the most difficult days of my long life. Both Christmas and New Year were spent on the boat. Once we had passed the English Channel and entered the Atlantic, the waves became impressive. Our third-class quarters were aft and low. The only deck to which we had access, the lower aft deck, would plunge down, almost to sea level, and then rise majestically to a height of several storeys. After a day, almost everyone was seasick, confined to their bunks, until the new continent was in sight.

In New York, whilst the Jewish charities looked for foster families, we were quartered in a Jewish orphanage on Amsterdam Avenue near 140th street, not far from Columbia University, which 15 years later, and for many years after that, was to become my host during some of my more productive years. On the one hand, the two months in the orphanage were not easy for

a)

b)

Fig. 1.2. Parents (**a**) Father and mother in Bad Kissingen, 1930 (**b**) Father, dressed as cantor, in front of the Tora Shrine, Synagogue of Bad Kissingen

children who had been spoiled by a reasonably normal family life; the food was not very pleasant, and we probably were also not used to the discipline of an institution. But everyone, inmates and staff, was most kind. What I still remember with pleasure was our reception in the local New York public school. The friendliness of the teachers, as well as the students, contrasted with the style of the old continent.[1] Learning English at 13 was no trouble at all. After not more than a month, we were perfectly at ease with the new language, much more so than I am now with French after a period of decades living in Geneva, at a riper age.

After about two months, we were sent to Chicago, 24 hours away by train in those days, to be inducted into our new home, with the Weissbrot family on

[1] This open acceptance of strangers and immigrants seems particular to the New Continent. This was impressed upon me a few years ago when visiting the Fermi National Laboratory near Chicago. I found a former close colleague in Geneva, a Pole, who had been working there for a year. He told me that he felt accepted by the community. For instance, he recounted what a pleasure it was for him when, moving into his house, the community "welcome wagon" came by. He considered himself American. Before that he had worked for several years in Switzerland as well as in France, but had never considered himself Swiss or French, nor would have been considered as such by the population.

Fig. 1.3. Jewish social life in pre-Nazi Germany. Left to right: Dr. Meyer, mother, Mrs. Meyer, father, dressed for a Purim (Jewish ersatz for carnival) party. My parents escaped to the U.S., the lives of the Meyers ended in the Theresienstadt concentration camp

the city's South Side. This did not work out, despite good intentions on both sides. It must be difficult in general to integrate adolescents into foster homes; and here, in addition, there were quite large cultural differences between the German Jewish community and the Eastern Jewish traditions that were more prevalent in the U.S. I don't remember the problems clearly, it was not an open conflict, but it didn't work out either. Perhaps one small matter that I do recall may give a hint of the problems: Each day we were given some change to buy lunch at school, but we tried instead to save these nickels and dimes to help make it possible some day for our parents to come over, or so we thought.

Fig. 1.4. Our family, 1926. Left to right: Hans Jakob (later Jack), mother Berta, father Ludwig, brothers Rudolf and Herbert

While waiting for new foster parents, we were quartered near Milwaukee, north of Chicago, in a beautiful old farm that had been turned into a summer camp for needy children. After a month or two, homes, this time separate, were found for us. These were in the rich northern Chicago suburbs along lake Michigan, not far from one another. My new foster parents were childless, in their mid-50s. Mr. Faroll ran the grain part, at the Chicago Board of Trade, and his brother the stock part, on Wall Street, of a joint brokerage business. The Winnetka home included Renoirs, as well as a cook and a chauffeur. In the fall of 1935, I was enrolled in New Trier Township High School, as was Herbert; once again we were in the same school class.

Coming back from the next summer's European trip, Mrs. Faroll was diagnosed with leukemia and died within weeks. Uncle Barnett, although now very much alone and without any experience with, or need for, children, bravely carried on with me in the house and did what he could to take care of me, although my life now was more with the servants than with the master. I am afraid that it was not a very rewarding task for him. Adolescents are not easy to assimilate, and in this case there was also the fact that I had come from a very different background, and at 14 had already fixed opinions. Perhaps the worst of these clashes was that already in those years I had come to believe that making an honest living meant doing honest work, and somehow stockbroking was not honest work. Although a socialistic outlook has stayed with me all of my life, my distinctions between good and bad have blurred. It is not obvious to me now that my own life's work, following my pleasure in learning about the laws of physics, while generously supported by society, was any more "honest" than Mr. Faroll's brokering.

Some of the weekends and a good part of the summers were spent on the Faroll Farm, near Aurora, some 60 miles' drive in the Lincoln eight-cylinder touring car with removable top, isinglass windows that could be strapped on, and extra seats that could fold down. For town travel there was the formal Lincoln town car with glass partition between the chauffeur and the master and a speaking tube for communication between them. The 300-acre farm was chiefly for growing corn and pigs. One of the two traditional farmhouses was used by the Danish farmer and his family who tended the farm on a crop-sharing basis, and another old house was at the disposition of the owner. There were two ponies and two riding horses, for Mr. Faroll and his guests, and for some time these were augmented by a racehorse breeding effort, which included half a dozen mares; an impressive stallion; a Mexican couple, the horse trainer and his wife; and in springtime perhaps two or three colts. During the summers, my job was to clean the stables of the horses other than the stallion; the latter was ferocious, which made it too risky for a somewhat undersized adolescent (Fig. 1.5).

The high school reflected the well-being of the region. It was nationally known for its excellent facilities and teachers. I had no problems in school, and for graduation made the honors class (I think), but I was not conspicuous for brilliance either. Normally I went there on my bicycle, but on rainy days sometimes got a lift from the chauffeur. The teachers and subjects I enjoyed most were mathematics, chemistry, and woodworking. In woodworking class I was able to build a substantial workbench, on which I then installed a carpentry shop in the Faroll basement, complete with small woodworking lathe, circular saw, band saw, and hand tools. For sports I joined the swimming team (New Trier had its own pool), but my skills were limited, and for the competitions I was never needed. To make a bit of pocket money, I delivered a local advertising newspaper once or twice a week.

Although this was now America, not Germany, anti-Semitism was no stranger to the scene. Mr. Faroll was prominent on the Chicago Board of Trade, and during my stay he was elected to be its vice-president. It was the custom to have the vice-president succeed to the presidency, a one-year job, the next year; but apparently his Jewishness prevented this, in what was something of a scandal. Jews were generally not accepted in social organizations of any pretension such as fraternities, athletic clubs, etc. On summer weekends,

Fig. 1.5. At 15, modelling my benefactor, Barnet Faroll, in Winnetka, IL

I earned a few dollars carrying golf bags, but at a Jewish country club. It was not impossible to become professor at a university, but a candidate's possibly Jewish origins were considered relevant by essentially all universities.

With the help of Mr. Faroll, my parents and younger brother obtained U.S. immigration visas, and toward the end of 1937 they also arrived and were installed on the Faroll farm, while I stayed in Winnetka until the summer of the next year, long enough to finish the term and to graduate from high school in 1938.

Father then was 63. His only job had been that of cantor and Hebrew religious teacher in Bad Kissingen, and no related work seemed possible in the new surroundings. It was still the Great Depression, and finding employment was in general very difficult, more so for a not-so-young immigrant who spoke little English. My parents were advised by the Jewish charities as well as by Mr. Faroll; and with their financial help, they bought a family business – a small "Jewish delicatessen" store in Rogers Park, the northeastern part of Chicago. An apartment was found around the corner, and I now rejoined the family. My parents had no experience in business nor any gift for attracting clientele, which soon diminished. But they never grudged the long hours (the store was open Saturday and Sunday as well as evenings), and many clients remained loyal just to help the immigrants survive, so that a minimal existence was possible. Mother had the chief responsibility for the accounts and the dealings with the suppliers. Father did the cleaning up, delivered orders on his bicycle, was the expert cutter of the lox (smoked salmon) traditionally served with bagels for Sunday brunch, and cooked the corned beef required for Saturdays. We all took turns working behind the counter. One big asset was the fact that mother spoke English without difficulty. Father's English was almost nonexistent when he arrived, but he had made the remarkable decision not to speak German anymore, even with the family; we only spoke English. Perhaps even more remarkable was the fact that he made no effort to maintain Jewish traditions, such as going to synagogue or praying. There were no outward signs of his Jewish faith, and I suspect that he may have rejected it. But the matter was never discussed, a fact that I have since regretted; it would be interesting to know how he really felt about his religion.

1.3 Higher Education and War

It was taken for granted that I would continue my education. My preference might have been medicine, but I was advised that this would require too much time and financial support to be realistic, and that it was necessary to choose a profession which would be likely to offer a job after a shorter study time. Engineering was considered reasonable. I managed to win a scholarship to Armour Institute of Technology in Chicago (then named after the meat-packing family, now renamed Illinois Institute of Technology) in a competitive examination, and so entered there in the fall of 1938 to study chemical engineering. From Rogers Park to Armour was 40 minutes on the elevated train. The store was at the foot of the "El," around the corner from the Rogers Park station, so that I could stay with the family and help with the store evenings and weekends. I was able to earn some pocket money by rolling nearby tennis courts. This gave me the chance to learn the game, which was a regular source of health and pleasure, until arthritis brought an end to this in my sixties.

At Armour I followed a course in elementary physics, but this has left no memory; I enjoyed particularly mathematics (becoming assistant to the very nice teacher, Prof. Oldenburger) and organic chemistry. There was a fine young English teacher, Prof. Hayakawa, who later became well-known and successful in San Francisco politics. I earned some money working in the lunchroom a few hours per week. There were fraternities, among which a separate one for the Jews. During the freshman year I became a suppliant "pledge" in the Jewish fraternity, but on termination of the pledge period I was rejected as unsuitable. I don't really know which character trait was the most unacceptable. The studies at Armour came to an end after two years, since there was no more scholarship money for me, despite a good academic record.

It was essential to find a job, but where? I visited the various chemical laboratories known to me to find work in some assistant capacity, filling out forms, but with no success. It was 1940, Europe was at war, but in the U.S. the depression was still strong. I don't know what would have happened to me if it had not been for a well-to-do businessman who had taken an interest in

Herbert, now studying English literature at Northwestern University. When he learned of my need, he contacted a pharmaceutical firm in Chicago, G.D. Searle and Co., who hired me to wash chemical glassware in their research labs at $13 per week, which was fine. One and a half years later the employment situation changed dramatically, with the American entry into the war following the Pearl Harbor attack. It took the war to get the U.S. out of a dozen years of great economic difficulty. It may be interesting to note that although all of the new industrial activity was aimed at the production of instruments for killing people, not very consumable, the social conditions of the bulk of the population improved dramatically.

The researcher to whom I was assigned did sugar chemistry, using extracts of pituitary glands, which were delivered once a week, perhaps by the Armour Packing Company. I helped him in the routine parts of his work, cleaned his glassware, learned some chemistry from him, and sometimes, after working hours, did some experiments to improve the work of my mentor. I also went to night school. Since Armour Institute did not have evening classes, my classes were with the University of Chicago, which offered chemistry but not chemical engineering. So my field of study changed to chemistry. After a year of this, in 1941, I was offered a scholarship by the University of Chicago, and so could return to daytime studies and finish a bachelor's degree in chemistry in 1942.

In the meantime, we were at war. Waiting for my draft call, I accepted work as assistant to Prof. Simon Fried, a physical chemist working on rare earth elements using spectroscopy. I found living quarters in a six-room house, cooperatively rented by about fifteen students, and also participated in the student eating co-op, a couple of blocks closer to campus on the same Ellis Avenue, where about fifty of us managed to take care of our eating needs. After the war, the Stagg Field athletic-field stands, a couple of hundred meters away on Ellis Avenue, became famous as the site of the experiments in which Fermi and coworkers had produced the first self-sustained nuclear chain reaction. This was going on just in those months in 1942, and at least two of my housing co-operative friends were actually participants in this history-making event. Of course, at the time they could not talk about this.

Then along came Colonel Sawyer of the U.S. Army Signal Corps. His daughter Tommy was one of the managers of our eating co-op. Probably not without his daughter's insight, he got the idea of establishing a program at the university, in electricity, magnetism and radio waves, to train young men who were then to be inducted into the Signal Corps to serve with the newly invented radar units, employed in antiaircraft defence. I signed up. So did Joe Heller, a most arresting young man who somehow appeared in our co-op circle at this time. He had us all in awed attention and admiration, as he told of his adventures in the International Brigade in the Spanish Civil War, all fiction as we learned later. I have often wondered if he was the same person as the later famous author of that name. In any case, I am most indebted to this Joe Heller. When he signed up in the Signal Corps office, he was dating Joan, a secretary of that office, and a few days later he introduced me to her, since in the meantime he had found someone who interested him more. Three months later Joan and I were married.

The courses were more advanced than the elementary physics course I had taken at Armour and were given mostly by young physics students no older than myself. They offered me my first look at Maxwell's equations, and, as much as any event, marked my transition from chemistry to physics. They occupied the fall of 1942, and their end was marked by my marriage to Joan, without ceremony. We offered a drink to half a dozen fellow students of the course and got one wedding present: a candy bar.

Once the course finished, I expected to be called to active duty. However, a small incident changed this, and the rest of my life as well. It is one of several chance happenings which had large effects on my future. As the course neared its end, I learned from one of the fellow students that he would be off to some research lab at MIT instead of active service. This seemed very nice to me, and I telephoned someone in the Signal Corps office to ask if I too might not be needed in this laboratory.

As a result, the beginning of 1943 found Joan and myself with a room in a lovely old house in the Boston Back Bay area, across the Charles River from the MIT Radiation Laboratory, the wartime laboratory for the development of radar bomb sights. The radar was used to locate targets in night bombing. The targets were often a city and the civilian population in it; I am not proud of my participation in the night bombing of civilians, but I am not sorry that the war was won by the Allies rather than by Germany and Japan.

The Radiation Laboratory had its roots in microwave advances in Great Britain just before the war, especially the invention of the multi-cavity magnetron, which could deliver microsecond pulses at high power levels. With some three thousand employees, it was large, even by present standards. Some of the physicists were young and brilliant, such as Julian Schwinger, who was admired for solving difficult mathematical boundary value problems, and who soon after the war became one of the founders of modern, renormalizable quantum electro dynamics. Ed Purcell headed the advanced development group, which spearheaded the move from 10 cm wavelength to 3 cm and eventually to 1 cm, and after the war became co-discoverer of nuclear magnetic resonance, the phenomenon at the base of what is now one of the most powerful diagnostic tools in medicine. Robert Dicke invented beautiful electronic tools, still important today, and later became one of the first to search for and understand the all-important cosmological microwave background radiation. Luis Alvarez invented an ingenious high angular resolution antenna that was to go into the leading edge of the wing of the B-24 bombers, for essentially its whole length; but for some reason it was never used. There were other excellent physicists, but many incompetent ones as well; myself a good example, with almost no training. There was not a sufficient number of trained physicists to deal with the technological challenges of the war.

I was assigned to the antenna group, about 30 people, including some trained physicists, a mathematician, several untrained physicists, technicians and secretaries, whose job it was to design and test the radar antennas as well as the radomes (the structures to house the antennas, which were necessarily located outside the bombers). The typical antenna was a shallow parabolic reflector, perhaps a meter in diameter, fed by a small horn at its focus. In order to achieve maximum resolution, the wavelengths used were the shortest that the available technology permitted. In the beginning this was the 10 cm band, later it was 3 cm. Antenna design was mostly a matter of making sure that the radiation pattern was not spoiled too much by "side lobes" or by the radome.

Perhaps my chief accomplishment at the Rad Lab was the design, late in the war, of a small receiving antenna to detect radar from Japanese war ships. The specifications called for something with large bandwidth and insensitivity to the radar's polarization. I produced a little horn, something like a flower opening out, in bronze. It received linearly polarized waves about equally well for all polarization directions, and so I thought that it was unpolarized. It was, in fact, accepted and produced in some small quantity. Only some weeks later I learned, from one of my colleagues in the antenna group with better physics training, the elementary, basic fact that all antennas are necessarily polarized, with a definite polarization, and will therefore not receive the opposite polarization. My antenna, which received equally well different linear polarization directions, must therefore have been circularly polarized in one sense, and was then insensitive to circular polarization in the opposite sense. I tell this little story only to illustrate that under these war conditions not all of the work was as competent as one could normally expect.

During these war years the regular physics teaching program at MIT was very limited. Many of the teachers, as well as most of the students, were engaged in war work or in the army. Nevertheless, it was possible for me to attend some courses. These were my first solid courses in physics: quantum mechanics and solid state physics, both taught by Prof. Lazlo Tisza, well-known for his work in atomic physics. I also enjoyed a course in Riemannian geometry by a renowned mathematician, Dirk Struik. These studies were a great privilege and after the war made it possible for me to enter graduate school in physics without serious handicap, despite the change in field from chemistry to physics.

In the fall of 1944, a son, Joseph, was born, and in the following fall I was inducted into active service in the army, since the war with Germany had been won and the work at the Radiation Laboratory was no longer of high priority. Joan and Joe found a home with Joan's mother and sister in northern Wisconsin. Perhaps the following anecdote from the six weeks of basic training is worth telling since it highlights my social inadequacies. The army was still segregated. We were given instruction regularly by a sergeant, who in one particular session explained the first aid kit issued to each soldier. The kit included some penicillin, one of the first antibiotics to be widely used. The sergeant made the point that it was important not to use this penicillin prematurely, as some "niggers" are wont to do to treat their syphilis. I saw fit to report this to the officer in charge, with the comment that I didn't think that we were fighting this war to promote racism. My reward was to be assigned, with two other misfits, to a very hard, disagreeable job (the digging of the sump pit) during the last two days' field exercises. During these six weeks of basic training, nuclear bombs exploded over Hiroshima and Nagasaki. The nuclear physics underlying this tragedy was totally unknown to me then.

After basic training, I was transferred to the New Equipment Introductory Division (NEID), in a New Jersey Signal Corps camp, Ft. Monmouth, which some years before had introduced the proximity fuse to the troops fighting in Europe. Two of the officers at Ft. Monmouth I came to know later at Columbia University, where we were together in the physics department for many years: Lieutenant Joaquin Luttinger, who had also studied with Lazlo Tisza and is remembered for pioneering work in theoretical atomic physics; and Captain Leon Lederman, with whom I later had the pleasure to collaborate on an experiment. In Ft. Monmouth, since I was a private and they were officers, we didn't know each other. From Ft. Monmouth I was transferred to be part of the support of an antiaircraft radar installation on some peninsula off Los Angeles. The war had come to an end. By accident I learned that soldiers with intention to return to studies had some priority to be discharged. I applied and was discharged after only seven months of active service.

1.4 Graduate School and Fermi

Early discharge was not the only way in which the government encouraged soldiers to return to studies; there was also, and very important in my case, the G. I. Bill, which provided financial assistance. Without this help it would have been much more difficult for our little family to manage my studies.

While I was still at MIT, a colleague had told us of the great Oppenheimer and advised me to try to do my graduate studies under him. I therefore applied to California Institute of Technology, where Oppenheimer had taught before heading the development of the atomic bomb at Los Alamos. Given my financial situation, I asked for a fellowship, but it was not granted. Because I had not known about assistantships (financial support which obligates the student to part-time teaching service), I had not asked for one. This might have been easier to obtain than a fellowship and would have been perfectly acceptable. However, given this failure, I applied to the University of Chicago, where my parents lived and where I had studied before. Chicago offered an assistantship. This accidental choice of institution turned out to be one of the most fortuitous and decisive events in my life. As it happened, Oppenheimer had accepted the directorship of the Institute of Advanced Study at Princeton, and so did not return to Caltech. On the other hand, Chicago had acquired a luminous physics faculty immediately after the war, which included Enrico Fermi.

I arrived at the University in the spring of 1946. The family was housed in one of several dozen small but very adequate prefabricated shacks that had been set up just south of the Midway for returning soldiers with families. While waiting for the fall semester to start, I found a job at the Metallurgical Laboratory on Ellis Avenue; the wartime center for the chemistry necessary for the bomb, especially that of plutonium. My bosses were Herb Anderson, associate of Fermi since his student days at Columbia University before the war, and Aaron Novick, later a biologist, a former housemate of the Ellis Ave. housing co-op, and during the war, a chemist at the Met. Lab. From these months at the Met. Lab. two happenings have stayed with me. One day, a small cylindrical capsule arrived, perhaps 4 cm in diameter and 15 cm long. I believe that it contained lithium hydride that had been irradiated in the Hanford, WA plutonium breeding reactors. It was introduced into a glass vacuum system and opened up; the gas contained in it was transferred by the use of Toeppler (mercury) hand pumps to a small glass bulb and sealed off. We then marched into a photographer's darkroom and admired the blue glow. The gas that had been collected was the first macroscopic amount of tritium – now so useful in luminous watch dials and in atom bombs – ever produced. The second incident also involved this tritium. It became known to Felix Bloch, coinventor of nuclear magnetic resonance, that this tritium existed, and he suggested that a small quantity be sent to him at Stanford University, where he could easily measure its nuclear magnetic moment with the nuclear magnetic resonance technique, of which he had been coinventor shortly before. But, instead, we set up an experiment ourselves to measure this not uninteresting quantity. This measurement required a considerable effort to set up apparatus which existed already at Stanford. In this way I learned the basic fact that scientists are not only interested in learning, but they are also, and perhaps not less so, motivated by ego, and the desire to be the first to publish something of interest.

The end of the war marked a major renewal of the University of Chicago physics department, led by the young Norwegian crystallographer William Zachariasen. In the pre-war years, the department had been chaired by K. T. Compton, of X-ray scattering fame, who seems to have been a more gifted physicist than administrator, and in the period just before the war the department was perhaps less than glorious. In addition to Fermi, Zachariasen succeeded in attracting Edward Teller, Maria Goeppert, and Joseph Edward Mayer, all from Los Alamos, as well as the field theoretician Gregor Wentzel from Zurich, solid state physicist Clarence Zener, and others. These, in turn, had attracted an excellent group of students, including Owen Chamberlain, Geoffrey Chew, Marvin Goldberger, T. D. Lee, Marshall Rosenbluth, Walter Selove, Rudi Sternheimer, Lincoln Wolfenstein, and Frank (C. N.) Yang. Of the dozen or so new graduate students, perhaps half had worked on the Los Alamos project and had followed their leaders to Chicago. I probably learned as much from these fellow students as from the professors; all were much ahead of me. Before coming to Chicago, Frank Yang already knew all the material

Fig. 1.6. Joan and Joe during graduate student days at the University of Chicago

that we covered in our courses from his studies under the most difficult circumstances, in a war-torn China devastated and brutalized by Japanese armies. Frank went on to make several outstanding contributions to our understanding of elementary particle theory. In 1954, together with Mills, he wrote the now-famous paper that is the foundation of non-Abelian gauge field theories, which are at the base of the present Standard Model of particle theory. In 1956, T. D. Lee, together with Frank, suggested ways in which one might look experimentally for the possible violation of left-right, or parity symmetry in particle interactions. In doing one of their suggested experiments, C. S. Wu discovered parity violation only some months later. Owen Chamberlain in 1956 was co-discoverer of the antiproton; Murph Goldberger, after an outstanding career in particle theory, became president of Caltech; Marshall Rosenbluth became a dominant figure in plasma physics; and Lincoln Wolfenstein has been an outstanding figure in our as-yet-unfinished search for a theory of the violation of the combined symmetry of charge conjugation and parity, CP symmetry, which was discovered in 1964 and which plays an essential role in the formation of our matter-antimatter asymmetric universe.

An exhilarating spirit of common interest in, and commitment to, physics reigned among the faculty and students. We worked together as one circle of friends. At the center was Fermi. He devoted a great deal of time to the students and taught many courses. In the fall of 1946, he gave the elementary physics course for undergraduates, and I could watch him as one of the assistants in the course. Within the same academic year we also had courses in electrodynamics and in nuclear physics from Fermi. In the evenings, I don't remember how often, perhaps once a week, he would offer some extracurricular discussions in physics. He would suggest a problem and invite us to come up with an answer. These sessions have been described by Yang in his "Selected Papers and Comments" [2], which also lists some topics. Examples include: the Thomas factor of 2, synchrotron radiation, and Johnson noise. The topics would be discussed the next time either by a student or by Fermi.

There were also not infrequent intimate seminars in Fermi's office, to which graduate students were invited, perhaps occasioned by the visit of an expert on a topic of especial interest. One of these, which I remember particularly, was by Schwinger, during the period of intense search, following its discovery, of a relativistically valid calculation of the Lamb Shift, and which successfully came to an end in 1948 with the works of Feynman, Schwinger, and Tomonaga.

Fermi made a great effort to be accessible to the students. If not otherwise occupied, he would be available to discuss our questions, and then the office door would be closed and he would be yours as long as necessary. This behavior was in contrast with that of Teller, who I remember was available only as long as someone more interesting to him did not pass the corridor. Fermi also took pleasure in having lunch with students. Several times a week we had the privilege to be with the great man around the lunch table in the Hutchins Commons student cafeteria.

Fermi's courses were beautiful. Simplicity was put above other considerations such as generality, so that the basic, underlying physics was made as clear as one could make it. In this connection, Marshall Rosenbluth has the following anecdote:

> While taking the course on nuclear physics, I found a definite but subtle mistake in a derivation concerning phase shifts. Naturally, with the arrogance of youth, I went to his office to show him his error. 'You are quite right, young man. However–' He went over to a huge set of files along the wall and pulled out a lengthy, original, mathematically sophisticated and rigorous derivation, with an appendix indicating the version suitable for a lecture, if not quite correct. He jokingly reassured me that his reputation as an experimentalist would be forever ruined if I were to leak out that he secretly indulged in such high powered analysis. I myself was not bright enough to notice these sacrifices of rigor in favor of pedagogic simplicity, but I appreciated the simplicity which he achieved.

In addition to the courses of Fermi, I remember with pleasure Teller's on quantum mechanics, Wentzel's on field theory, and that of Zachariasen on optics (Fig. 1.7).

My first contact with ongoing physics was a seminar early in 1947, in which Fermi discussed the result of a cosmic-ray experiment performed in Rome by Conversi, Pancini, and Piccioni [3]. In 1935, Yukawa postulated a particle of intermediate mass, about 200 times heavier than the electron and one-tenth as heavy as the nucleon, to explain the spatial range of the forces which bind nucleons in nuclei. In 1937, Stevenson and Street, and independently, Neddermeyer and Anderson, using cloud chambers, discovered particles of such mass in the cosmic radiation. It was only natural to identify these "mesotrons" with the Yukawa particles. Through the work of Rossi and Rasetti, using Geiger-counter arrays, it was learned that the mesotron is unstable and decays with electron emission, with a lifetime of about two microseconds. The new experiment, using similar Geiger-counter techniques, looked for decay electrons of mesotrons stopped alternatively in sheets of carbon or iron. The new feature of the experiment was the selection of either electrically positively or negatively charged particles using a simple, ingenious magnetic deflection.

As expected, the positive mesotrons, which are repelled by the electric fields of the atomic nuclei, decayed freely in both materials. However, it was expected that negative mesotrons should quickly find themselves in an atomic orbit very close to the positively charged nucleus. If the mesotron had the strong coupling needed for Yukawa nuclear binding, the subsequent nuclear capture times would be much faster than the decay time. Consequently, as Fermi put it, the mesotron would be "eaten up by the nucleus before it had time to die in bed." This was indeed the case for iron, but in carbon the negative mesotrons decayed quite happily. This showed that the nuclear interaction of what is now called the muon is much too weak to play the role of nuclear binding envisioned by Yukawa.

With this seminar, particle physics entered my life. The real Yukawa particle, the pi meson, was found [4] in Bristol only a few months later. Toward the end of the same year, Rochester and Butler [5] in the laboratory of Blackett in Manchester, in a cloud chamber triggered on showers produced by cosmic rays in a layer of material above it, found the first two examples of decays of yet another type of new unstable particle, later called strange particles. 1947 was a good year for particles.

The new department rules required the students to take a "basic" examination as a prerequisite to thesis study. Of the dozen of us who took this in the summer of 1947, all but one passed. The rules specified that this exam could not be repeated; Teller had the unpleasant job of informing me. Perhaps he was selected because I had worked under him on a problem that he had suggested: the analysis of some experiments that had been performed at nearby Argonne Laboratory on the scattering of slow neutrons in magnetized iron.[2] Teller advised me to drop physics in favor of something for which I might be better suited. Somehow, however, I was permitted to take the exam again, a few months later, and was told that I did very well. Why the rules were dis-

[2] Independently, Gian Carlo Wick was working, not far away, on the same topic. He had come from Rome after the war to be a professor at Notre Dame. On one of his visits to his longtime friend, Fermi, we discovered that we were working on the same topic and finished the work and published it together. This was the beginning of a lifelong friendship.

Fig. 1.7. Snapshots from graduate student days: **a)** Enrico Fermi; **b)** Gregor Wentzel; **c)** Lazlo Tisza and Edward Teller; **d)** Frank Yang; **e)** between sessions at an APS meeting (left to right: Geoff Chew, John Reitz, Lincoln Wolfenstein, Bill Rarita, Murph Goldberger, and Sid Dancoff)

obeyed I don't remember. Probably Zachariasen, the department chairman, who was also a kind man, found it hard to show the door to an ex-G.I. with a three-year-old child.

For the thesis I wanted to do something theoretical. Fermi had accepted me as a student, but a suitable topic was not obvious. The clear theoretical challenge of the day, following the discovery of the Lamb Shift, was the calculation of radiative corrections in quantum electrodynamics, despite the fact that naively these were infinite. Fermi was very interested in this, but did not consider himself nor his students, and certainly not me, a match for the challenge. At one point, he asked me to look into a problem mentioned in a paper by Rossi and Sands on stopping cosmic-ray muons. These had been identified, as usual at the time, by the observation of the decay electron some microseconds later, but there were too few of these by a factor of about four. I thought that I could understand about half of this missing factor on geometrical grounds (this was before the advent of computers, and geometrical acceptances were not always trivial to calculate), but there was still a missing factor of two. It had been assumed that the energy of the decay electron, as expected for the simplest two-body decay, was one-half of the rest energy of the muon. Fermi had the idea that maybe this was the problem, since the energy of the decay electron had never been measured. He asked me to write the authors to suggest that they measure the electron spectrum. When they replied that they had not the time, Fermi suggested that I do this as my thesis.

The experiment measured the flux of the decay electrons as a function of the thickness of material that they could penetrate, that is, their range. There was a setup and an attached lady, Mrs. Woods, for filling Geiger-counters with the proper gas, and there was expertise, which helped in the design of the coincidence electronics. For the rest, I was on my own. The experiment that I designed (Fig. 1.8) was a bit more ambitious than the norm at the time. There were 80 counters instead of the more usual dozen; these were 2.5 cm in diameter and 50 cm long. More challenging, some of the brass tubes needed to be turned down from the commercially available wall thickness of 0.4 mm to one-half of that, but this the department machinists were able to do. Construction required perhaps three months.

When first turned on, of course, as is usual, it didn't work. There were far too many decay electrons. This was in the morning, and I told Fermi. When we went to lunch with three or four fellow students, all theorists of course, Fermi was very pleased to tell them about my problem, to show them that experiments may also have challenges. Fermi has, of course, made great contributions of a theoretical nature, but he also enjoyed doing experiments, and some of the early experimental studies on neutrons were by his group in Rome before the war, and, later, he experimented with pions at the Chicago cyclotron. The problem was understood in the afternoon. It turned out to be trivial: a ringing electronic circuit.

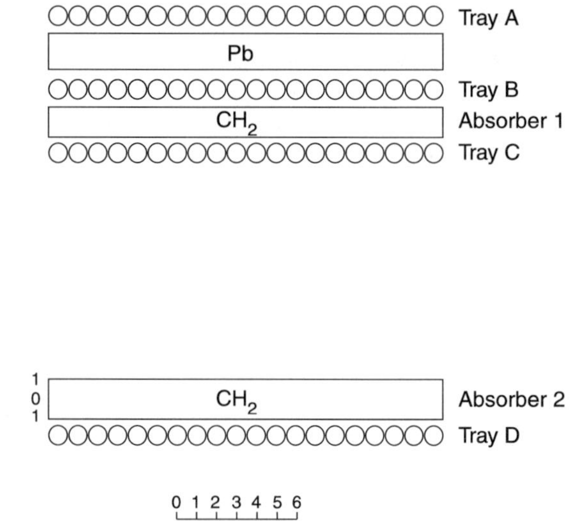

Fig. 1.8. Apparatus for my thesis experiment to measure the range distribution of muon decay electrons

The first results were published in June 1948. They showed substantially smaller electron energy than one half of the muon rest energy. In order to increase the statistical accuracy, it was decided to take the experiment to the top of Mount Evans in Colorado, the highest mountain in the U.S. with a road to the top, more than 4,000 m high, where the muon flux is three times that at sea level. A half-ton truck was rented, the experiment mounted inside, and a young man engaged to drive it, since I did not know how. On top there were some primitive facilities such as a diesel-driven electricity generator and sleeping quarters, as well as a restaurant for the tourists. There we spent two months. With the improved data it was possible to show [6] that the electron spectrum is continuous (Fig. 1.9), therefore requiring two neutral particles, probably neutrinos, in addition to the electron in the decay, and that therefore the muon was a spin 1/2 particle, a fermion very much like the electron, but 200 times as heavy.

Fermi supported the experiment, but did not offer advice, except at the end, in the analysis of the data, when he cautioned not to forget to take radiation into account in the calculation relating the electron's energy to its range, to keep me from making what would have been a serious error. I think that he deliberately restrained himself from directing me, since he probably could have done many things much better than what he saw me doing. In leaving me alone, he probably helped me a great deal.

The result [6] was of some importance. Although I was not clever enough, several people [7–9] immediately independently recognized that the rate of the three-body muon decay, the muon capture rate measured by Conversi et al. [3],

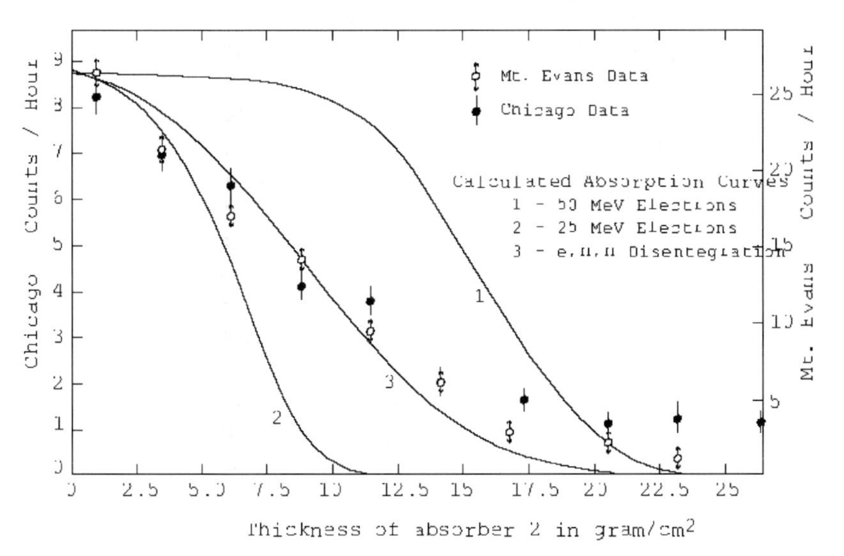

Fig. 1.9. The observed range spectrum which showed that the energy spectrum of the electron in muon decay is a continuum

and the beta decay rates could be understood in terms of a generalization of the very successful interaction Fermi had proposed in 1934 for nuclear beta decay, called then the "Universal Fermi Interaction." This represented a big step forward toward the present electroweak theory.

In connection with the discovery of the Universal Fermi Interaction, there is an interesting story involving Bruno Pontecorvo, a fascinating, inventive figure in the history of particle physics. In 1947, soon after the discovery by Conversi et al. [3], he published a letter in the *Physical Review* [10], pointing out the equivalence of the nuclear capture rates of negative muons and of electrons if allowance is made for the difference in the sizes of their atomic orbits. This represented a first smell of the Universal Fermi Interaction, two years before it was later accepted. However, Pontecorvo's beautiful suggestion was universally ignored, even by Fermi, who had been Pontecorvo's teacher and was his friend. Fermi was, of course, highly interested in this physics and surely had read Pontecorvo's letter. I believe that this shows that new ideas are not always easy to accept, sometimes even by the brightest and most open of people such as Fermi. I know that at the time I had difficulty in accepting the notion that one can make a parallel between the muon and the electron.

Some remarks about Fermi as we students knew him may be of interest. He was simple, direct, confident, without complication. He had no need to be reassured about his greatness, professional or otherwise. He had little interest in artistic or humanistic culture. Apart from his relation with family and friends, he concentrated on physics, and that was enough. Once we passed a painting in which the artist had painted the sky green. Fermi observed that he cannot see any reason why an artist should paint the sky other than blue. In parlor games, at his house or elsewhere, he was interested in winning, whether at musical chairs or pitching pennies. He often used his bicycle to come to work and played tennis, but I never had the chance to play with him. I do not remember any discussions on politics, even at lunch. They were times of intense questioning, within the government and by physicists, concerning the future of the new and dangerous nuclear weapons; I know that Fermi was involved in these discussions, but he never saw a need to discuss these questions with us.

He really loved to understand physics, and if some new idea interested him, he would insist on understanding it clearly. He would work through the problem in his own way, systematically, and write the result in a notebook, a diary of his work. These notebooks he kept on shelves above his desk and have been preserved. If he later needed to refresh himself on the question, he would know how to find it in his notes. A notebook was always with him, as was his six-inch slide rule. He was clear and organized. Often, in later years, I regretted that I did not try harder to follow his luminous example; but even the most perfect teacher can lead the less gifted only so far.

Fermi was interested in understanding concrete questions of physics, but less so in more philosophical questions. I doubt that the recent concern in quantum mechanics with the relationship of the observer to the experiment would have interested him greatly. What mattered was to be able to calculate, to understand, and to solve the problem. He was, of course, a genius in theoretical physics, but he also enjoyed experimenting. I think that this was in part due to the fact that experiments can be welcome relaxation compared to the intellectual pressure of trying to understand something which you don't understand. While I was doing my thesis experiment in his laboratory, he was building there, in those days before computers, a little analogue electronic device to integrate the one-dimensional Schrödinger equation. A few years later, Fermi and I were doing similar experiments at the new cyclotrons, he at Chicago and I at Columbia, and I was able to admire the virtuosity with which Fermi did the experiment as well as the analysis. Fermi was superb not only as theoretician, but also as experimenter.

2

Institute for Advanced Study, 1948–1949, Theory

After completion of the thesis, I became free to return to my interest in theoretical work. This time I succeeded in becoming a disciple of Oppenheimer, whom I had failed to find for the doctorate. The fall of 1948 found our family comfortably installed in temporary housing quarters set up after the war for visitors at the Institute for Advanced Study on the outskirts of Princeton. The Institute staff was divided between very distinguished permanent members and younger, generally less distinguished members with appointments for a year or so. It was an especially privileged place where people were free to do research with no other duties. There were no laboratories, although a few years later, von Neumann established a laboratory for electronic computer research, before there was commercial interest in electronic computers. The Institute was beautifully situated, within a short walk of the Princeton University campus, but outside the city. Behind was a vast, swampy Urwald that extended to the end of the world.

The senior faculty included some of the most celebrated intellects of the time. The dominant schools were mathematics, with, among others, Goedel, Weyl, and von Neumann, and physics, with Einstein, Oppenheimer, Placzek, and the young Pais. Uhlenbeck and Yukawa were senior visitors. Illustrious figures in other fields included the art historian Erwin Panofsky and the political scientist George Kennan. I must admit that there was not a great deal of contact between the young postdoc and the illustrious figures in other disciplines.

Oppenheimer was director of the Institute and guru of physics. At the same time, he was the dominant advisor to the government on matters concerning nuclear weapons. A few years later, following the famous Oppenheimer Hearings,[1] he was excluded from defense consultation, but at the time, to get to

[1] These were supposed to be secret, but were published almost immediately. I recommend them to the young reader for a taste of the McCarthy political climate in the U.S. at the time.

Oppie's office, you had to go through an ante-office with secretary, security guard, and a big safe for secret documents.

We were about a dozen postdocs that fall: Ken Case, Freeman Dyson, Dan Freer, Bob Karplus, Bruria Kaufmann, Norman Kroll, Joe Leport, Hal Lewis, Cecile Morette, Sheila Powers, Fritz Rohrlich, and Ken Watson. We had close relations among ourselves and a common interest in the physics of elementary particles. This was shared by Oppenheimer, Yukawa, and Pais. Placzek worked on neutron and radiation diffusion, and although we had no physics contact, we often had lunch together. He was a remarkable person, one of the most interesting in my life. Uhlenbeck had by that time strayed from atomic physics, I believe to statistical mechanics. I remember the seminars he gave that year as extraordinarily beautiful and clear. Einstein did not interact with us. He had quarters in another wing of the building and talked with his secretary, we were told. Once he came to a seminar. It was given by his friend of pre-Hitler times, Max von Laue, one of the unfortunately not too many German physicists of that time who distanced themselves from the Nazis. I never had the courage to talk to Einstein. It would not have been easy. I knew very little general relativity: What could we have discussed?

When I wrote that we were a dozen postdocs, this was not quite true. Freeman Dyson was no "doc." Although he had done more than enough at Cornell to get such a degree, he felt, as he told us, that his Oxford Masters was good enough, or maybe better, so he preferred to do without a Cornell Ph.D. Freeman was the youngest, but also the most brilliant and the unquestionable star of that year. During 1948, both Schwinger and Feynman had come forward with versions of quantum electrodynamics that were both manifestly relativistically covariant and capable of giving finite answers in radiative correction calculations, such as the Lamb Shift. However, their formulations were very different, and both were difficult for lesser humans to understand; in the beginning, their equivalence was only understood by Dyson. As I remember it, his first and great accomplishment on coming to the Institute was to make Feynman and Schwinger, and their equivalence, accessible to the rest of the profession. During the remainder of the year, he made fundamental contributions to the theory of getting finite results in these perturbation calculations to any order in the perturbation expansion. It must have been an especially rewarding year for Freeman.

For me the year was both gratifying and frustrating. In November, we became four with a second boy. It was a special pleasure to get to know J. A. Wheeler and J. Tiomno of Princeton University, who had used my experimental results on muon decay in their paper proposing the Universal Fermi Interaction. Dyson became a family friend; he was still a bachelor and often at our table. He is one of the most extraordinary, intelligent, independent thinkers I have known. Also, the opportunity to get to know Oppenheimer, Uhlenbeck, Yukawa, Pais, Placzek, Norman Kroll, Ken Case, Ken Watson, and others was a privilege (Fig. 2.1). In fact, part of my pleasure in physics, especially during those early years, was in the appreciation of the qualities of some of my colleagues and the pleasure in being with them. My own performance in theory was less glorious. Most of the year went by without it being possible for me to latch on to some problem I could solve. This was the frustrating part.

In the spring of 1949, luck came my way in the form of a topic. I noticed that some perturbation calculation results in particle physics (as distinguished from quantum electrodynamics) that were infinite could be made finite with the help of a regularization scheme that had just been invented by Pauli and Villars [11]. In this way it was possible to calculate the lifetime for a spin-zero neutral pion decaying into two photons [12]. Although the neutral counterpart to the charged pion had not yet been discovered, it was rather plausible to expect this particle to exist. It had been pointed out by Kemmer in 1938 that the observed charge symmetry of nuclear forces could be understood if the Yukawa particle exhibited isotopic spin symmetry, which requires a neutral counterpart to the observed positively and negatively charged particles. Also, Oppenheimer had speculated on the existence of a neutral meson to explain the early development of electromagnetic showers in the propagation of cosmic rays through the atmosphere. The calculation also made use of a suggestion by Fermi and Yang, that the pion is a composite of nucleon and anti-nucleon [13]. This is still the picture if we substitute quarks for nucleons.[2]

I remember Oppie's pleasure when he saw the paper, not because it concerned a particle he had proposed, but because he was relieved that something had come out of me before the end of my year at the Institute. I believe that the paper was not uninteresting at the time, both from the point of the method used as well as the result. In more recent times it has found some echoes in connection with the discovery of the so-called "anomalies" in modern particle theory [14].

Oppenheimer was not only Institute director, he was also spiritual leader of the part of the physics group concerned with elementary particles. As Fermi met with his students in the evenings, Oppie had us in his office maybe once a week. Of course, the topics were at a higher level. He himself was not directly working on any of the physics, but was deeply interested and very quick in following a new topic. He was "spiritual" director in more than one sense. The Oppenheimers resided in a lovely mansion just a hundred meters from the main building and hosted us not infrequently. The walls were adorned

[2] When I did this calculation in 1949, I was unaware of the fact that the calculation, to begin with, is not well-defined. Only in the late 1960s (Adler, Bell, Jackiw) was it realized that there are other ways to regularize the expression, giving other results. It is now known under the name "anomaly." The most remarkable feature is that the symmetry properties of the underlying equations are not maintained in the result. Depending on the regularization method that is chosen, one or another symmetry is lost. The correct method is to maintain the gauge symmetry and, for the case I calculated, the axial vector current conservation is lost. By chance, my selection of the Pauli-Villars regularization did this. The anomalies now play a fascinating, important part in modern gauge theory.

Fig. 2.1. Snapshots at the Institute for Advanced Study, 1949: **a**) George Uhleubeck, co-discoverer of the electron spin, **b**) Freeman Dyson, **c**) Abraham Pais and the Yukawas, **d**) contemporary postdocs. Left to right: Ken Watson, Dan Freer, Ken Case, Joe Leport, Norman Kroll, Hal Lewis, and Fritz Rohrlich

with the family art collection, which included famous impressionists, and we were initiated not only into the right way of doing physics, but also the more correct way of drinking, such as dry martinis. Joan and I became friends of the Oppenheimers; I respected him a great deal, and this seems to have been mutual at least in part because he gave my name to *Time* magazine as one of ten promising young physicists. I am afraid, however, that he overestimated my cultural level, because at one moment he read to me from the Bhagavad Gita.

Fermi and Oppenheimer were exceptional persons as well as physicists; they were also very different. Oppenheimer had a broad need of immersion in human culture quite apart from physics and a deep sense of social responsibility. As a teacher, he probably had an even greater impact on the development of American theoretical physics than Fermi. As a graduate student, I used mimeographed lecture notes of courses Oppie had given before the war. He was considerably less rigorous in his need to understand a question thoroughly and more vulnerable in his need for reassurance.

3

Berkeley, 1949–1950, Accelerators

Early in 1949, while on a visit at the Institute, Gian Carlo Wick (Fig. 3.1) asked if I would come to Berkeley the next year as his assistant. Two years before, when I was graduate student at Chicago and Wick was professor at nearby Notre Dame University, we had worked briefly together on a problem concerning the scattering of slow neutrons. Now, he was working on the new quantum electrodynamics (QED); probably the purpose of his visit to the Institute had been to discuss this with Dyson. Every physics student now learns (QED) using some beautiful mathematical methods invented by Wick at that time. It was easy to accept his offer. In addition to its excellent physics department, the University of California at Berkeley hosted the Radiation Laboratory,[1] created and directed by E. O. Lawrence, and then the largest accelerator center in the world. Its 340 MeV,[2] 184-inch cyclotron was, in 1948, the only accelerator with sufficient energy to produce the newly discovered mesons.

Early in the summer, the four of us took the train to California: three days and nights in Pullman comfort. I still remember, on arrival, paying my respects to the department chairman, R. T. Birge, who was also known for keeping the measured values of the light velocity, electron mass, and fine structure constant up to date, in the days before Art Rosenfeld took over with the Particle Data Book. Birge was a bit shocked to see me. Apparently he did not expect me before fall and was in difficulty with the department budget, which was late in coming. He informed his young visitor more or less as follows: "Here you are already. We expected you in the fall. But I don't have my budget yet, how will I pay you? A more thoughtful person would not

[1] Sorry about the two radiation laboratories, but the two radiations were quite different, those at MIT were at an energy of 10^{-6} eV, those at Berkeley at 10^{+8} eV.

[2] The MeV, one million electron volts, is a unit of particle energy. 1 eV is the energy acquired by a particle with unit electric charge, that is the charge of the electron, in traversing a potential difference of one volt.

Fig. 3.1. Gian Carlo Wick (with permission by Jack Steinberger)

have come so unexpectedly." This happening aside, our welcome in Berkeley was cordial. Apart from Gian Carlo, former fellow Chicago graduate students Geoff Chew, Murph Goldberger, and Owen Chamberlain were there to greet us. Shortly after, Murph left for MIT (I hope that he didn't arrive too early!) and sold me his old car, my first.

Lawrence, who together with Livingston had invented the cyclotron, had much stronger financial support from the government than any other physicist at the time. He had built up a formidable laboratory with a strong staff and unmatched facilities and technical support. During the war it had worked on the production of trans-uranic elements as well as the separation of uranium 235 using "calutron" mass spectrometers, both in connection with the nuclear bomb development. The Rad Lab was the center for particle physics activity at Berkeley, theory as well as experiment. Lawrence was in charge. The seminar hall, which I remember as an old wooden building, had in the front row, in line with the plain wooden chairs, a pretentious leather armchair, in which, if attending, he would place himself; otherwise it would remain empty. By the time I had got there, Lawrence no longer had a strong connection with the ongoing physics and de facto had left the physics direction of the laboratory to Luis Alvarez. The spirit of the laboratory reflected its unique capabilities; it did not lack self-confidence or self-appreciation, to the point of being arrogant. It was the world, the rest was of minor significance. It was common not to bother to publish in the journals, a Rad Lab report was sufficient and perhaps more prestigious. The 184″ had come into operation in 1946 with 380 MeV alpha particles. In 1947, the pion was discovered in Bristol and, in 1948, with the help of Lattes, a young Brazilian participant in the Bristol experiment, it was seen in nuclear emulsions at the cyclotron, still operating with alphas. When I arrived in the summer of 1949, the cyclotron was operating with 340 MeV protons, and there was a good deal of experimentation with pions such as measurement of its mass and the determination of production cross sections (production probability) of pions on different materials, using chiefly nuclear emulsions as detectors. In retrospect, probably the most interesting experiment in progress was that of Bjorkland et al. [15], which, with a rudimentary pair spectrometer,[3] studied the spectra of gamma rays (photons) emitted when the protons struck a target, both forwards and backwards with respect to the proton direction, and at different proton energies. The observed rates and spectra were very difficult to understand unless one assumed that the gammas were the decay products of a neutral counterpart of the pion.

Although I was hired as Wick's assistant, presumably to do theory, and did manage a small piece of phenomenology with Geoffrey Chew during the year, once at Berkeley I was attracted by the experimental possibilities at the Rad Lab, and Gian Carlo generously supported this. At first, I imagined using the magnetic cloud chamber of Wilson Powell to make a more precise measurement of the spectrum of the muon decay electron, rather crudely determined in my thesis. The muons were to be stopped in a small liquid scintillator in the middle of the chamber, and its expansion was to be triggered by the signal produced in the scintillator by the stopped muon and its decay electron, a few microseconds later. Although I worked on this for some months, the experiment was not done; I have no memory of the reason.

At about this time, the 330 MeV electron synchrotron, invented, designed, and built by Edwin McMillan (Fig. 3.2), began to be operational. The electrons were never brought out of the machine, but at an internal target they produced a beam of photons with bremsstrahlung spectrum, that is, a continuous energy spectrum up to the electron energy. No one was using the machine, people were used to the cyclotron. I no longer remember the circumstances, but Ed, who was not interested in doing an experiment himself, encouraged me to think of doing something with his device. I owe the excitement and success of the remainder of that year to the constant encouragement of McMillan. I began to consider measuring the production of mesons by the photon beam.

At the time, Robert Hofstadter, then professor at Princeton University and some years later famous for the first measurement at Stanford of the electric charge "structures" of the proton and neutron, was visiting. He happened to be one of the first to be familiar with the newly discovered use as particle detectors of the organic scintillating crystals, anthracene and stylbene. When electrically charged particles traverse these materials, flourescent light is produced that can be converted into an electronic signal by photomultiplier devices. The latter had been developed for automobile headlight dimming, but had never been used for this. Hofstadter introduced me to this technique.

Somebody on the Hill (local slang for the Rad Lab) was able to grow the crystals; the electronics I soldered together myself. Graduate student A. S. Bishop and I were soon installed in the photon beam of the McMil-

[3] In a pair spectrometer, the energy of a gamma ray is measured by converting the gamma ray into an electron and a positron, and measuring the energies of each of these on the basis of their deflection in a magnetic field.

Fig. 3.2. Edward McMillan (with permission by Jack Steinberger)

lan synchrotron. In the little experiment, the photons were allowed to strike a target cylinder, 5 cm in diameter and 5 cm long, alternately carbon or polyethylene. By taking the difference of the rates on the two targets, the rate for hydrogen, or protons, could be deduced. Some 20 cm away, on a frame that could be rotated about the target and beam direction, an aluminium absorber was placed whose thickness could be varied in order to select the energy of the observed meson. This was followed by three anthracene crystals (see Fig. 3.3). The electronics was arranged to detect positive pions coming to rest in the second crystal. These would decay into a muon in a time shorter than the reaction time of the electronics. The decay muon, with a range of only 2 mm, would generally stop and decay in the same crystal. The time-delayed muon-decay electron was observed in crystals 2 and 3. The experiment could not detect negative pions that are swallowed up by the atomic nucleus before they can decay. A similar counting technique was developed by Alvarez and used to measure the muon lifetime [16].

The detector worked just fine and permitted the measurement of the cross section (probability) for the production of pions by photons on carbon and hydrogen as a function of the pion production angle and its energy [17]. At the time, some of the most fundamental properties of the pion, such as its spin and parity,[4] were still not known. It was imagined that the spin might be zero, the simplest possibility and the one assumed by Yukawa. The observed angular distribution (see Fig. 3.4) was inconsistent with the dipole

[4] The basic properties of elementary (that is, not composite) particles are their mass, spin, parity, and interactions with other particles. The spin is the particle's angular momentum in units of the Planck constant divided by 2π, and the parity is its property under space reflection, either positive or negative.

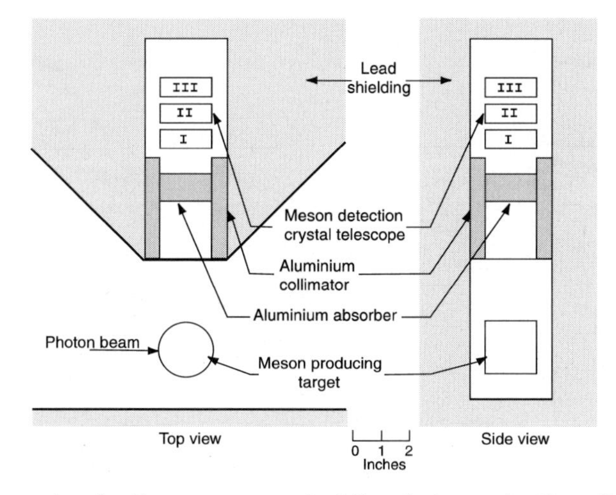

Fig. 3.3. Apparatus for the measurement of the photo production of positive pions

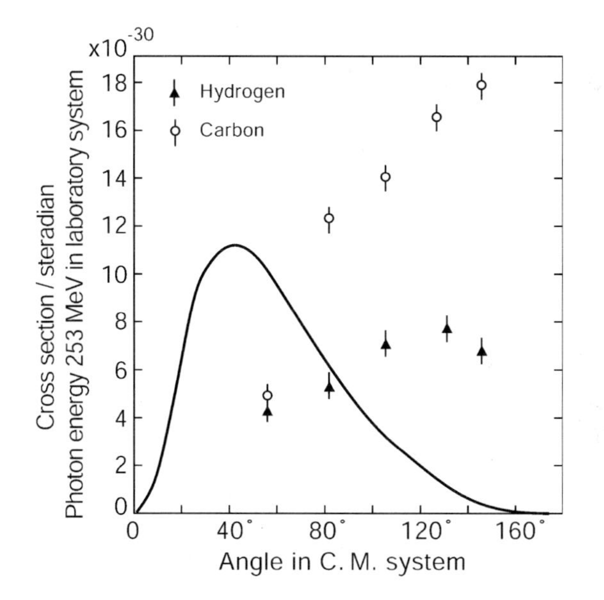

Fig. 3.4. Angular distribution of the photoproduced mesons. This disagreed with the expectation for a scalar pion (solid line), and so constituted first evidence for the pseudoscalar nature of the pion

distribution expected for scalar (spin zero, even under reflection) mesons and represented first experimental evidence for the pseudoscalar (spin zero, odd under reflection) nature of the pion.

Very shortly after, I was joined by W. K. H. Panofsky (well-known for the experiments performed some months later on the production of neutral pions by negative pions coming to rest in hydrogen and for the development in the

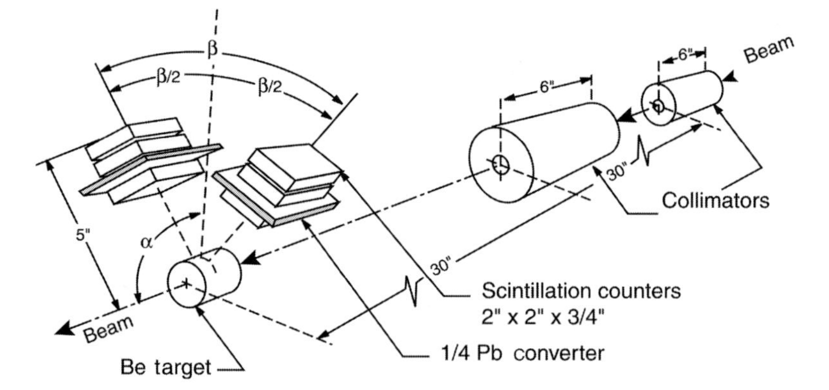

Fig. 3.5. Apparatus for the detection of the photons in the decay of a neutral pion to photons

1960s of the Stanford two-mile-long linear accelerator) and graduate student J. Steller to try, with similar technology, to detect the photons from a possible two-photon decay of a possible neutral pion, whose existence had already been indicated by the experiment of Bjorklund et al. [15] referred to above, in which single photons were detected from a target struck by protons.

The detector, which a few years ago could be seen as an appendix to the McMillan synchrotron exhibited at the Smithsonian Museum in Washington, consisted now of two telescopes, one for each of the gamma rays, and each with three scintillation crystals. The first scintillator, in anti coincidence, was followed by a lead plate a few mm thick in which the photon was to convert into an electron–positron pair, and the final two crystals were in coincidence to detect the pair. The angle between the telescopes could be varied and so could the angle between the plane of the telescopes and the incident photon beam, as can be seen in Fig. 3.5. We expected the two photons to come out in opposite directions, so we started out with the telescopes at 180° to each other. After an effort of several days, we managed to reduce the background sufficiently to achieve a clear signal from the target. To confirm the pionic decay origin of the signal, we then decreased the angle between the telescopes to 90° and expected the signal to go away. Instead of going away, the signal increased by a factor of ten! It took a day to figure out that we were seeing the neutral-pion decay alright, but the pion was moving with non-negligible velocity, and so, depending on this velocity and on the angle of this velocity with respect to the decay photons, a distribution in the angle between the two photons had to be expected. For the typical energies of the pions produced in this experiment, an angle near 90° was the most likely. This decay angular distribution was then measured for three production angles (see Fig. 3.6), and this permitted not only a demonstration of the existence of a neutral pion that decays into two photons, but also some rough conclusions about its production by photons and about its mass [18].

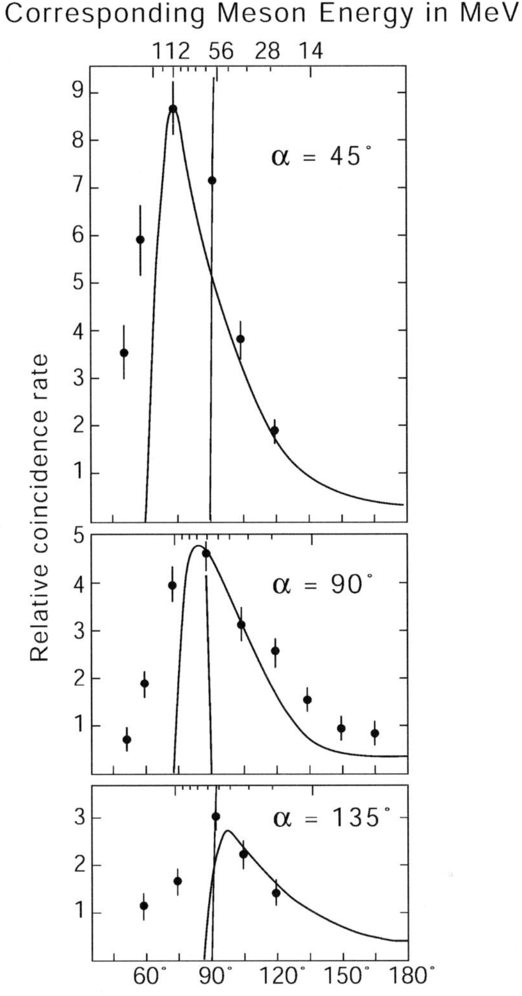

Fig. 3.6. Distribution in the angle between the two photons for three production angles. The points are the measurements, the lines are the expected angular distributions for pions decaying into two photons, assuming that the mass of the neutral pion is the same as that of its charged counterpart

On completion of this experiment, I was joined by my old friend and fellow student at Chicago, Owen Chamberlain, and his friend, Clyde Wiegand, who some years later, together with Emilio Segre and Tom Ypsilantis, demonstrated the existence of the antiproton, and also by Bob Mozeley in using the detector for photo-produced positive pions in a measurement of the pion lifetime, which was then only very roughly known. Clyde was an expert in what at the time was fast electronics; also the anthracene crystals were replaced by

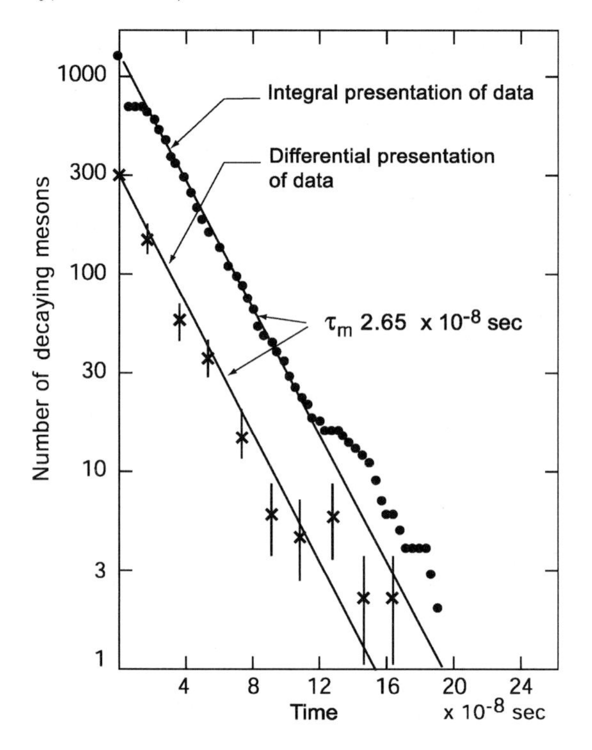

Fig. 3.7. Measurement of the pion mean life

stilbene, with faster light emission. Both the pion and the decay muon signal were displayed on an oscilloscope trace and photographed, so the time interval between the pion arrival and the muon appearance could be measured (see Fig. 3.7). The pion mean life was found to be $(2.63 \pm 0.12)^x\ 10^{-8}$ s [19]; the present value is $(2.6033 \pm 0.0005)^x\ 10^{-8}$ s.

This was 50 years ago. It was possible to have the pleasure of finding three quite interesting new results in one year, not counting one false start, together with one or two colleagues. Particle physics experiments have become much, much more difficult. Now, five years is a short time for the planning and construction of an experiment, and as many or more years may be required for accumulating and analyzing data, and the team of physicists typically numbers several hundred. It must be amazing to the younger generation that one could do several incisive experiments in a few months. The main reason, of course, is that in the intervening years we have learned quite a bit, and although there is no lack of remaining, unanswered, fundamental questions in particle physics, experimental progress is infinitely harder. Another essential element in the great satisfaction that this work offered was the uniqueness of the resources and consequent opportunities that the Rad Lab offered at the time. It is an advantage to be in the right place at the right time. This was the positive and overriding aspect of that year for me.

But there was also a negative side to that year, intimately linked to the cold-war political climate in the U.S. in general and at Berkeley in particular, and which was then in an early and virulent state. America prides itself on political freedom and individual liberties, but the facts at the time[5] were different. During that academic year, the regents of the University of California, a state institution, established the "Non-Communist Oath." The faculty was invited, on the threat of dismissal (which was however not explicitly stated), to sign a statement to the effect that they had never belonged to nor were presently members of the communist party, nor of a long list of organizations declared "subversive" by the U.S. Attorney General. It is interesting to note that this was an initiative of the State of California and not directly linked to the House "Un-American Activities" Committee, and so illustrates that the political intolerance was widespread; it was also long before McCarthy chaired the Senate committee. The physics faculty was divided into supporters of the oath, who included Lawrence and Alvarez; those opposed on the grounds that it violated political freedom, who included Wick, Serber, Panofsky, and Chew; and others, including the department chairman Birge, without strong opinion either way but who regretted the unpleasant divisions and animosities produced by the oath.

A lot of time was spent in discussions concerning the oath; I remember being lectured by Alvarez on the evils of the fifth column of communist "sympathizers," in whose traitorous ranks he probably included also me. Perhaps the following incident is also connected. Shortly after conclusion of the experiment with Panofsky and Steller at the McMillan synchrotron, in which the neutral pions were observed, I got permission to take the detector to the 184″ cyclotron in order to measure the production of these neutral pions by protons. However, only hours after the experiment was set up, I was asked by Alvarez to leave. No reason was given, so it is only a guess that this hostile act was related to political differences. The same hostility prevented approval of my nomination by McMillan and Segre to a regular faculty position. I had made it clear that I did not wish to stay otherwise.

Perhaps the level of acrimony is illustrated by my last day at the University of California–Berkeley. Since I had no further appointment at the university, I had accepted a position at Columbia University in New York, where a comparable cyclotron was under construction, for the academic year starting in the fall. Until then, I expected to continue working at the Rad Lab during the summer. However, on the 30th of June there was a note on my desk, stating that since I had not signed the oath, I was no longer welcome at the lab and must leave within the same day!

Most colleagues signed the oath. Some, for instance, Panofsky and Serber, signed with misgivings, in order not to damage an outstanding center of physics, but left a few years later. Several, including Geoffrey Chew, Gian Carlo Wick, Hal Lewis, and Howard Wilcox, declined to sign. They lost their

[5] as now (2003), given the "war on terror."

appointments at this most prestigious institution in the beautiful State of California and accepted perhaps lesser institutions and surroundings in order to maintain their self-respect. It was a large sacrifice, and I have not ceased to admire them for this. As best I know, none had ever been communists or members of organizations on the attorney general's list. My own refusal to sign was an easy gesture, since I had no proper job at Berkeley anyway.

4

Properties of Pi Mesons

What did I do during the summer of 1950 after my summary ejection from the Berkeley Radiation Laboratory? In fact, I don't remember. I imagine that I wrote of my predicament to Polycarp Kush, then head of the Columbia University physics department, to ask if he could accept me a couple of months in advance and that he agreed more graciously than Birge had managed to do the year before.

Columbia University, in the heart of New York City, has a distinguished place in American higher education. In particular, the physics department at the time had been elevated by Isidor Rabi to one of the three or four outstanding centers of physics in America. In the 1930s, Rabi had developed his method of measuring atomic spectra using atomic beams to select a particular spin state. The basic idea is still used, with important refinements. Rabi had attracted an outstanding school of atomic physicists to Columbia. In this field, the department was without parallel. There were Polycarp Kush, who had already discovered and measured precisely the anomalous moment of the electron; Willis Lamb, who in 1947 discovered the Lamb shift, that is, the splitting of the 2s-2p levels of the hydrogen atom; Norman Ramsey, who was to discover an important extension of Rabi's atomic beam method; and Charles Townes, who in 1954 invented the maser, the forerunner of the laser. But Rabi had also appreciated that the study of elementary particles physics was becoming of fundamental interest. He encouraged the expansion of the department in this direction. A 380 MeV cyclotron (the Berkeley one was 340 MeV) had already been built and had just started operation when I arrived. Gilberto Bernardini had been invited as visiting professor; my own appointment was the idea of Rabi. Particle physics got another big boost when two excellent theorists, Robert Serber and T. D. Lee, joined the department a couple of years later. Once again, I enjoyed the privilege of excellent surroundings. This was of basic importance to my work. Quite apart from the technical facilities such as the accelerator, which were of course essential, the possibility of discussion and the exchange of ideas with first-class colleagues have always been vital to me. The teaching duties consisted of one course per semester, in my case often atomic physics and elementary quantum mechanics, as well as possible odd jobs, which for me tended to be the German exam, since reading knowledge of German was required for the Ph.D. For the rest I was free to do research. The job, an assistant professorship, offered tenure from the start, a matter which was not even mentioned when I was offered the job. This can be contrasted with more recent university employment practices. It is now considered normal to employ young Ph.D.s for several years as postdocs, with low salaries and no job security whatever, then as instructors, still without job security. Tenure, if this is finally offered, comes after they have been working for eight or ten years, and are nearing 40.

There was no room for the massive cyclotron on Columbia's crowded New York City campus. Instead, the laboratory was located at Nevis, a magnificent estate 30 kilometers to the north, along the Hudson River, owned by the university. On it was an impressive mansion house built at the end of the 18th century by Alexander Hamilton, one of the founding fathers of the then newly independent United States. A good part of my next 15 years was spent in these beautiful surroundings. The laboratory personnel, including half a dozen professors and a dozen or two students, numbered perhaps one hundred. The drafting office was located in the mansion house, as were the living quarters of the Bernardinis. The purchasing office was in the old carriage house, the machine shop in the adjacent stables. Much of the estate was used as botanical gardens and a botany laboratory.

When I arrived in the summer of 1950, the Nevis cyclotron was the world's best source of pions until, a year later, when the Chicago cyclotron, with ~ 50 MeV more energetic protons, became operational. There was still much to learn about these new particles. My own interests centered on their basic properties: spin, parity, and interaction with other particles, neutrons and protons in particular. By the time I had arrived, a beam of charged mesons, external to the cyclotron and suitable for experimentation, existed already, unique in the world. This had been the work of Leon Lederman, then a graduate student, five years before my captain in the army, and John Tinlot, who had helped me in the summer of 1948 when I suffered altitude effects on arriving at the top of Mount Evans to do my cosmic-ray thesis. In the meantime, he had joined the Columbia faculty. The mesons were produced by allowing the accelerated protons to strike a target inside the cyclotron. The same magnetic field that keeps the protons circulating in the cyclotron deflects the charged mesons that are produced, some of them toward the outside, and one only has to cut a path through the thick concrete shielding to use them in an experiment. The momentum of the pions is determined by the trajectory that has been chosen. The lifetime of the pions, which we had measured a few months before, corresponded to a mean flight path of about 10 m, so that about one-half survived for experimentation outside of the shielding. A year later, these beams were improved considerably by the trivial device of a magnetic lens.

The experimental techniques I had used in Berkeley, based on scintillation counters, were immediately applicable here and much easier and more effec-

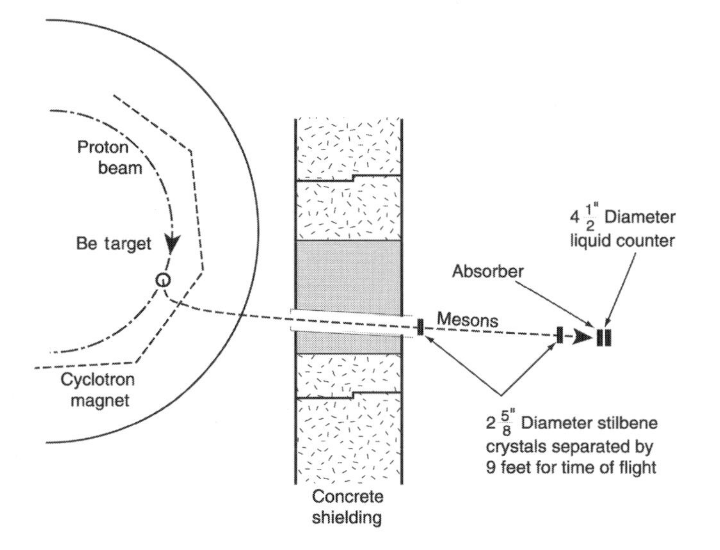

Fig. 4.1. Experimental arrangement for the first measurement of pion total cross-sections [20]

tive in tackling the problems that interested me than the nuclear emulsion and cloud chamber techniques that were then in use at Nevis. The first experiment was a measurement of the interaction probability ("cross section" is the technical term, the cross section of a process being defined as the number of interactions per beam particle, per target particle per unit area) of pions in various materials. This fundamental property of the pion, its interaction with nuclei, had never before been measured. It was done in collaboration with Alan Sachs, another fledgling professor and for many years after a much appreciated collaborator and friend, as well as student P. Isaacs and technician C. Chedester [20]. It was the all-time simplest experiment of my life. There were two small scintillation counters some 2 m apart along the line of the beam to define the incident pion. By measurement of the time of flight these permitted the rejection of the electrons in the beam, which move faster, and of the protons, which are slower. This was followed some 25 cm along the same line by a third counter slightly larger in diameter (see Fig. 4.1). A slab of the material whose interaction cross section was to be measured was inserted between the last two counters. Results for a proton target were obtained by subtracting the carbon rates from those for polyethylene. A three-fold coincidence represented an unscattered pion, but if the third counter showed no signal the event was classified as a pion interaction. For eight heavier elements, from lithium to lead, the measured cross sections were close to the nuclear area, $\pi \times A^{1/3}R^2$, with A the atomic number and $R = 1.4 \times 10^{-13}$ cm, known previously from proton scattering experiments on nuclei; but for hydrogen it was smaller, by a factor of four. This was of some interest because it could be put in relationship with the pion parity, that is, its reflection symmetry. This

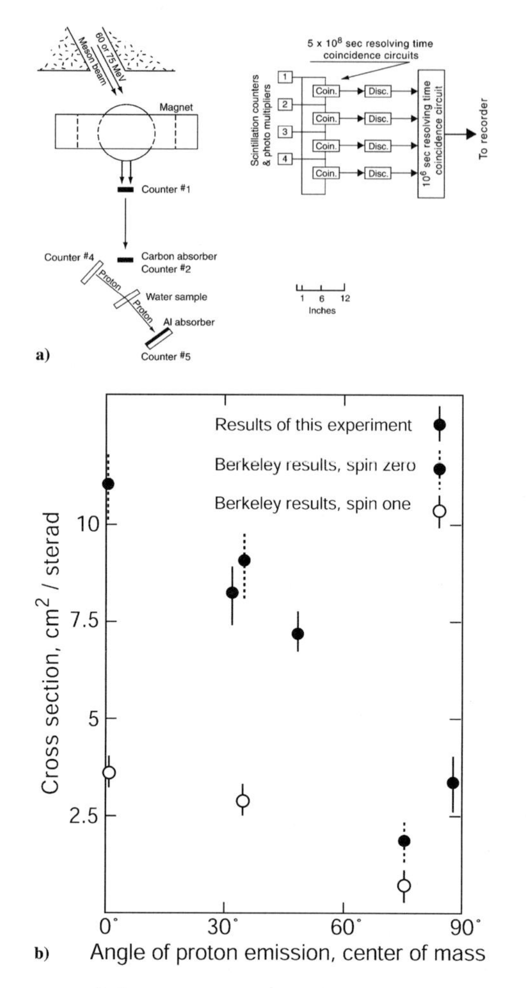

Fig. 4.2. Measurement of the reaction $\pi^+ + d \rightarrow p + p$ in order to determine the pion spin [23]. **a**) Experimental arrangement, and **b**) crosses and solid error bars are the results at three different center-of-mass angles, crosses with dotted error bars are Berkeley results for the inverse reaction on the basis of zero spin, circles with solid error bars on basis of spin one

small result could be understood as corroborative evidence for a negative pion parity, which already had been indicated by the photoproduction experiments [17]. The rather small energy of the pions in the experiment corresponded to wavelengths larger than the range of nuclear forces, or the radius of the proton, and a negative pion parity could be expected to inhibit the interaction rate in proportion to the square of the ratios of these lengths.

The spin of the pion, clearly a fundamental quantity, was not yet known, although when various experiments were put together, such as the two photon decay of the neutral pion and the observations on the capture of stopped negative pions in deuterium, spin zero was suggested. At about this time, Marshak [21] noticed that the spin of the pion could be directly measured by comparing the rate for the process 'proton + proton $\rightarrow$ deuteron + pion', with that of its inverse: 'pion + deuteron $\rightarrow$ proton + proton'. In this ratio the incalculable term which depends on the dynamics of the interaction cancels and only a well-known, kinematic contribution remains, which contains a factor equal to twice the spin of the pion plus 1: 1 if the spin is zero, 3 if it is one, etc. The forward reaction rate had already been measured in Berkeley [22]. The backward reaction, at the same center-of-mass energy, could be measured in the Nevis meson beam. The protons were detected in two scintillation counters, nearly back to back with respect to the target, and whose axis could be rotated with respect to the beam direction. The target was alternately heavy and light water (see Fig. 4.2a). The deuteron rate is just the difference in the two observed rates, since the reaction cannot proceed on hydrogen. Comparing the observed reaction rate with that of the inverse reaction measured in Berkeley, the spin of the pion was found to be zero [23] (see Fig. 4.2b).

At Nevis, in order to measure reaction rates on protons, we were obliged in these first experiments to subtract results using alternatively polyethylene and carbon targets. At the Berkeley Radiation Laboratory, technology in general and cryogenic technique in this example were more advanced. It had been possible to use liquid hydrogen already when I was there in 1949–1950. At a certain moment we learned that, in fact, a primitive, pre-war plant for liquefying hydrogen existed also at Columbia, just outside the Pupin physics building. It consisted of an atmospheric storage tank for the hydrogen gas, with a few cubic meters capacity, a compressor, a catalytic purifier, and a Joule-Thompson valve. Alan Sachs and I put this back into some kind of working order. A target was designed and built in which, in a thin-walled stainless-steel cylinder surrounded by vacuum, a few hundred cc of hydrogen could be exposed to the meson beam with as little matter in between as possible. Alternatively, deuterium could be put into the cup by condensing deuterium gas using the liquid-hydrogen reservoir (the boiling point of deuterium is a few degrees above that of hydrogen). It should be appreciated that we were not cryogenic experts. An experiment consisted of getting things ready at Nevis, then driving the 30 km along the Hudson River to the Pupin laboratory, getting the liquefier going and running it for a few hours to produce some 7 or 8 liters of liquid stored in a dewar (thermos bottle), putting this into the car, and driving back up the West Side Highway, and then running the experiment another few hours.

In those days, it was useful to be jack of all trades. Since then, the field has become dramatically more specialized. An experimental team nowadays may consist of 100 or more physicists, divided into several groups, each responsible for some specific aspect of the experiment, such as the construction of a particular part of the apparatus, or the design of data readout hardware, or for writing software, or for data analysis, etc. One of the interesting inadequacies of the work then was the matter of safety.

No one was responsible or competent or even thought much about it. An example is putting dewars of hydrogen into a car and driving off on public roads. I can remember having at least four accidents involving inflammable gases, which only by luck had no terrible consequences: Once the hydrogen liquefier blocked, once I pierced the vacuum of the hydrogen target with a screwdriver and the hydrogen boiled off within seconds into the room, once some plumbing which I had constructed on a hydrogen pressure vessel burst with a very big bang, and once 100 liters of liquid propane were released into an experimental hall. The question of safety received more attention following an accident with liquid hydrogen at MIT, in which one person died.

Safety at Nevis, as well as ease of liquid hydrogen experimentation, improved dramatically in 1953, with the purchase and installation of a Collins refrigerator offered to us by its inventor, Professor S. C. Collins of MIT. This little machine could make a few liters of hydrogen per hour using *helium* gas as refrigerant so that it was no longer necessary to compress and circulate hydrogen at high pressure. The helium was compressed and then cooled itself as it drove a turbine.

The first Nevis experiment using the liquid hydrogen target was the search for another decay mode of the neutral pion, its decay to photon, electron, and positron [24]. The π^0s are produced when π^- mesons come to rest in hydrogen and are captured by a proton. About one half of the time, the pion-proton atom disintegrates with the emission of a neutron and a π^0, the remainder of the time with emission of a neutron and a photon [25]. We were able to observe this decay, with a branching ratio of about 1/80. When we did this experiment, we were unaware that already one year before Dalitz had noticed [26] that, more generally, on the basis of the well established quantum theory of electrodynamics, so-called internal conversion of a photon to an electron–positron pair must be expected. For all reactions in which photons of sufficient energy are produced there are corollary reactions, with probability 1/160 that an electron–positron pair appears instead of the photon. These are now called Dalitz pairs. We, therefore, should have expected a branching ratio of $2 \times (1/160)$, which is what we found. Dalitz could do more: He also could predict the internal constitution of the pair, that is, the distribution in the angle between the electron and positron, and the distribution in the division of the energy. This experiment would not have been possible without the liquid hydrogen target, since the background from the conversion of the photons of the normal decay in heavier material would have been insurmountable.

A related experiment was the first measurement of the difference between the masses of the charged and neutral pion [27]. This could be obtained from the angular correlation of the two photons (see Fig. 4.3). The π^0, produced by π^- capture on protons as in the previous experiment, is emitted with a certain kinetic energy, essentially the mass difference times the velocity of light squared. As a consequence, the angle between the photons is not 180°,

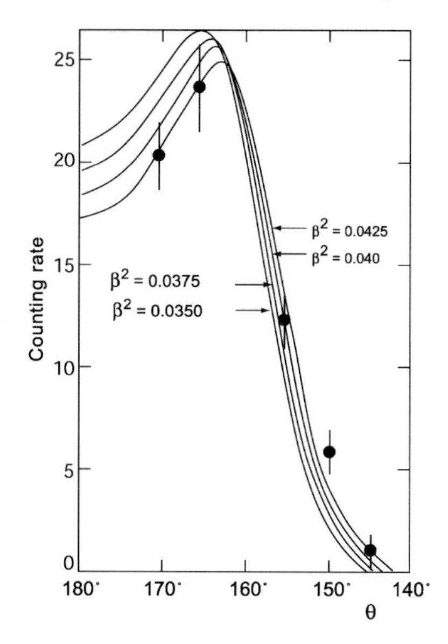

Fig. 4.3. Distribution in the angle between the two photons from the decay of the π^0 produced in the capture of a π^- stopped in hydrogen [27]. The typical angle is about 160°. This corresponds to a velocity squared of the π^0 of 0.039 compared to that of the light velocity squared, and a difference in the masses of charged and neutral pions of about 3.2%

but proportionately less, and so can furnish a measure of this mass difference. The charged pion turned out to be 3.2% heavier than the neutral, an effect which is attributed to its interaction with the electromagnetic field, absent in the case of its neutral partner.

One of the most important results of these early years of the study of the pion was the discovery of the so-called 3/2, 3/2 resonance in pion-nucleon scattering. 3/2, 3/2 refers to the spin and isotopic spin.[1] It was the first example of hadronic resonances. In the following years, dozens of other examples were discovered: mesons, baryons, un-strange and strange resonances. At the time, all had the same right to be considered elementary particles, as had the pion and the nucleon. All these particles are *hadrons*. They interact *strongly* with one another, as distinguished from electrons and neutrinos, known as *leptons*, which do not interact strongly. Since 1969, of course, we have known

[1] Isotopic spin is the name of a symmetry in which the members of a multiplet have similar properties except the electric charge. The number of members of the multiplet is 2 × (isospin) + 1. The neutron and proton are an isospin doublet with isospin 1/2, just as the electron is a *spin doublet* with two polarization states. The pion has isospin 1 and, consequently, is a *triplet* with three charge states: $+, 0, -$. The multiplet of the resonance with isospin 3/2 consists of four states with electric charges $++, +, 0$ and $-$.

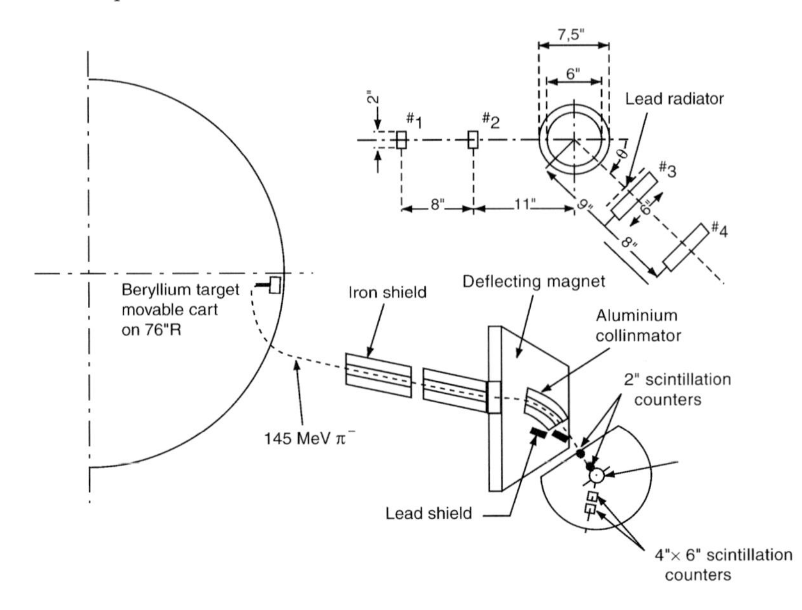

Fig. 4.4. Experimental setup for the measurement of the pion-proton scattering angular distributions in the experiment of Anderson, Fermi, Martin, and Nagle [28]

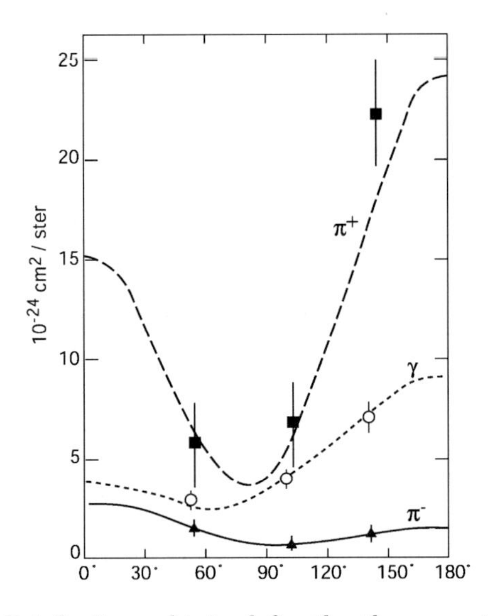

Fig. 4.5. Angular distributions obtained for the three reactions at 120 MeV by Anderson et al. [28]. The curves are the results of a phase shift fit to the data

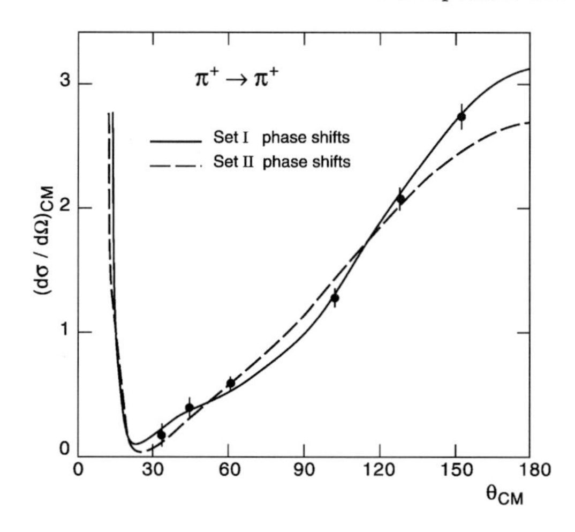

Fig. 4.6. Angular distribution obtained by Bodansky et al. [29] for the scattering of π^+ on hydrogen, at 60 MeV

that hadrons are not elementary, that hadrons are composites, composed of quarks. The resonances are related to the stable, or only weakly decaying, lower mass hadronic states composed of the same quarks; just as in atoms, the excited states are related to the ground states. The groundwork for the understanding of this resonance was done by Fermi and coworkers at the slightly higher energy Chicago cyclotron, which began to function in 1951. It was more definitively established by Ashkin and coworkers, using a yet slightly higher energy cyclotron at Carnegie Institute of Technology, which came into operation in 1953. Unfortunately, we at Nevis could only contribute marginally, because our protons, and consequently our meson beams, were of too low an energy. In particular, our positive pion beam was limited to about 65 MeV. At Chicago useful beams were possible up to 135 MeV, and at Carnegie up to about 200 MeV. Since the resonance energy was 180 MeV, we at Nevis were too far away. Fermi could almost reach it and Ashkin could finally see it clearly. In the three laboratories the experimental apparatus and procedure were nearly identical. Pion beams of either charge were incident on a liquid hydrogen target, and the three reaction rates, $\pi^- + p \to \pi^- + p$, $\pi^- + p \to \pi^0 + n$, and $\pi^+ + p \to \pi^+ + p$, were measured as function of the angle of the outgoing pion. Figure 4.4 shows the Chicago arrangement and Fig. 4.5 the three angular distributions measured at the pion energy of 135 MeV [28]. The π^+ results at our much lower energy of 60 MeV are shown in Fig. 4.6 [29]. Finally, the Carnegie Tech angular distributions at 170 MeV [30], near the top of the resonance, are shown in Fig. 4.7, and the energy dependence over a large energy scale [31] in Fig. 4.8. On the basis of the height of the peak of Fig. 4.8 and the angular distributions of Figs. 4.5 and 4.7, it was possible to assign the spin 3/2 to this isospin 3/2 resonance with the four charge

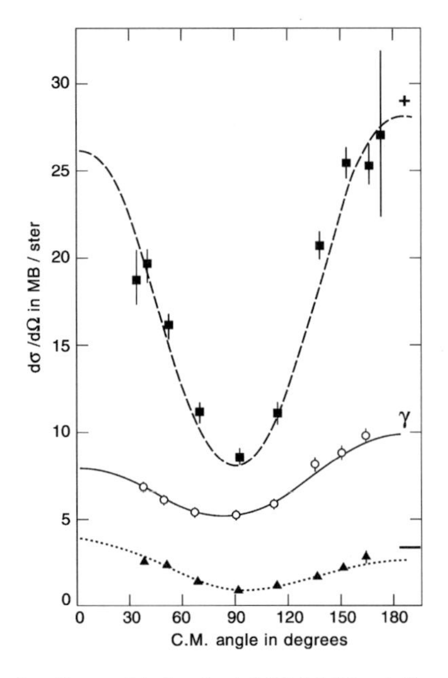

Fig. 4.7. Angular distributions obtained at 170 MeV, at the peak of the resonance, by Ashkin et al. [30]. The symmetry about 90° is characteristic of resonance

states: ++ and 0, experimentally observed; and + and −, not observed in these experiments.

A special pleasure for me in those days was the chance to watch Fermi, this time not as student, but with the critical eyes of a colleague doing similar work. It seems that from the beginning he had realized that the study of the pion-nucleon system required the measurement of all three reactions, over as wide an energy spectrum as possible. This understanding of the problem came much more slowly to me. The underlying scattering theory was well-known, but was not without its complexities. None of us, including Fermi, had ever worked on related problems, and Fermi showed the way, very correctly and also conservatively. He must have been excited about the indications of the resonance, but he never presumed this as long as there was an alternative possibility, as was the case. He obviously enjoyed doing this experiment, although most of his previous great contributions had been theoretical.

An example of his experimental ingenuity was the Fermi target cart. In those days, to change the pion energy, it was necessary to change the position of the target, inside the cyclotron, on which the protons produced the pions. This in turn required opening the vacuum chamber, and following the operation, a loss of time of several hours to re-establish the vacuum. Fermi got the idea of making a little four-wheeled cart to carry and position the target without opening the vacuum. The wheels rolled on the cyclotron magnet pole-face, following grooves that existed already for the purpose of achieving the

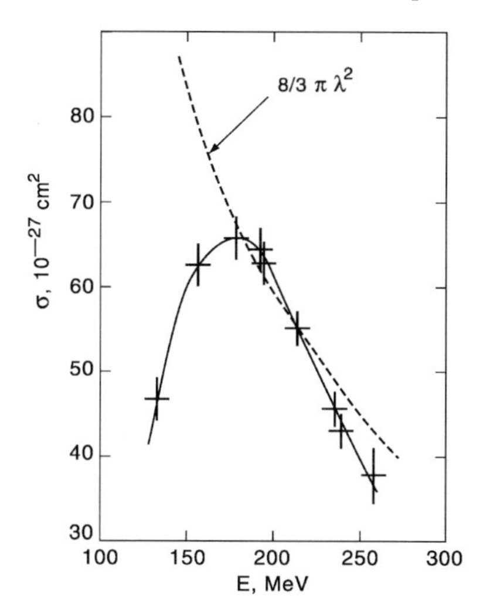

Fig. 4.8. Total cross section for negative mesons as function of energy [31]. The peak is characteristic of a resonance, and the magnitude shows the spin to be 3/2

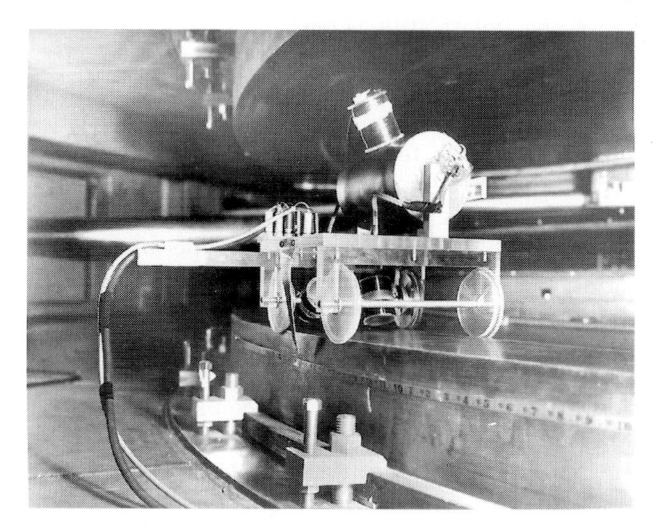

Fig. 4.9. Fermi cart (with permission by Jack Steinberger)

required magnetic field. A small coil was attached to one of the wheel axles. This, in the cyclotron's proper magnetic field, served as the motor to move the cart and target to the wanted position. Another coil was attached to an arm that held the target, and so could swing the target into and out of the beam. It was a lovely, clever device, and among us, only Fermi could have all these original ideas (Fig. 4.9).

These experiments left little room to doubt the odd parity of the pion, which for that matter was already rather clear. Nevertheless, I will mention here two experiments concerning the pion parity that were more direct. The first is the observation of the reaction in which a negative pion comes to rest in liquid deuterium and two neutrons are emitted: $\pi^- + d \rightarrow n + n$ [32]. This reaction is forbidden for even parity mesons, and so its observation shows that the parity is odd. The second experiment, which I remember with a good deal of pleasure, is the direct measurement of the π^0 parity. It was the doctor's thesis of Nicolas Samios, later co-discoverer of the Ω^- hyperon – which established flavour SU(3) symmetry – and even later, very successful and longtime director of Brookhaven National Laboratory. The experiment became possible only after the development of bubble chambers, and in particular bubble chambers filled with liquid hydrogen. It required the observation of the rare decay of the π^0 into two electrons and two positrons, that is, the decay in which both photons are internally converted to electron–positron Dalitz pairs. Each pair is typically emitted with an angle between the electron and positron of about 5 to 10 degrees. The normal to the place of an electron–positron pair reflects the polarization of the photon which gave rise to it and the distribution in the angle between the two normals is a direct measure of the pion parity, since for even parity the photons are emitted with parallel polarization, while for odd parity the relative polarization is perpendicular. The detailed theory had been worked out by Kroll and Wada [33]. The measurement of this correlation would have been impossible with the counting techniques available at the time. Even now it would be a challenge, but with the bubble chamber it became possible. In 1958, our 12″ (30 cm) hydrogen bubble chamber, which had been constructed to measure strange particle properties at the Brookhaven Cosmotron, became obsolete there when a larger chamber became operational, so it was brought back home to Nevis for this experiment. When a negative pion stops in the chamber it is captured by a proton and, as we have seen, half of the time this produces a π^0, which with the probability of $(1/160)^2 = 1/25{,}000$ decays into two Dalitz pairs. Of course, 1/25,000 is not very large, but with about 10 stopping pions per picture, and many pictures, some 100 such events were accumulated, sufficient for the measurement [34]. One of the events is reproduced in Fig. 4.10.

In the same picture one can see several other stopping pions, but without associated tracks, due to events in which either the π^0 decays, as it does normally, into two photons (invisible because they are not charged), or a photon is emitted instead of the π^0. As can be seen in Fig. 4.10, the distribution in the angle between the two plane normals agrees with the expectation for odd parity.

As a final example of these years of work on properties of pions, let me recall the charged pion decay into electron and neutrino. From the days of the pion discovery it was known that the electron decay of the pion, if it exists, is rare compared to that into muon. In the Universal Fermi Interaction theory there were two possibilities, depending on the form of the Fermi current which was assumed: Either the electron decay should be four times more probable than the muon decay, which was clearly not the case, or it should be 8,000

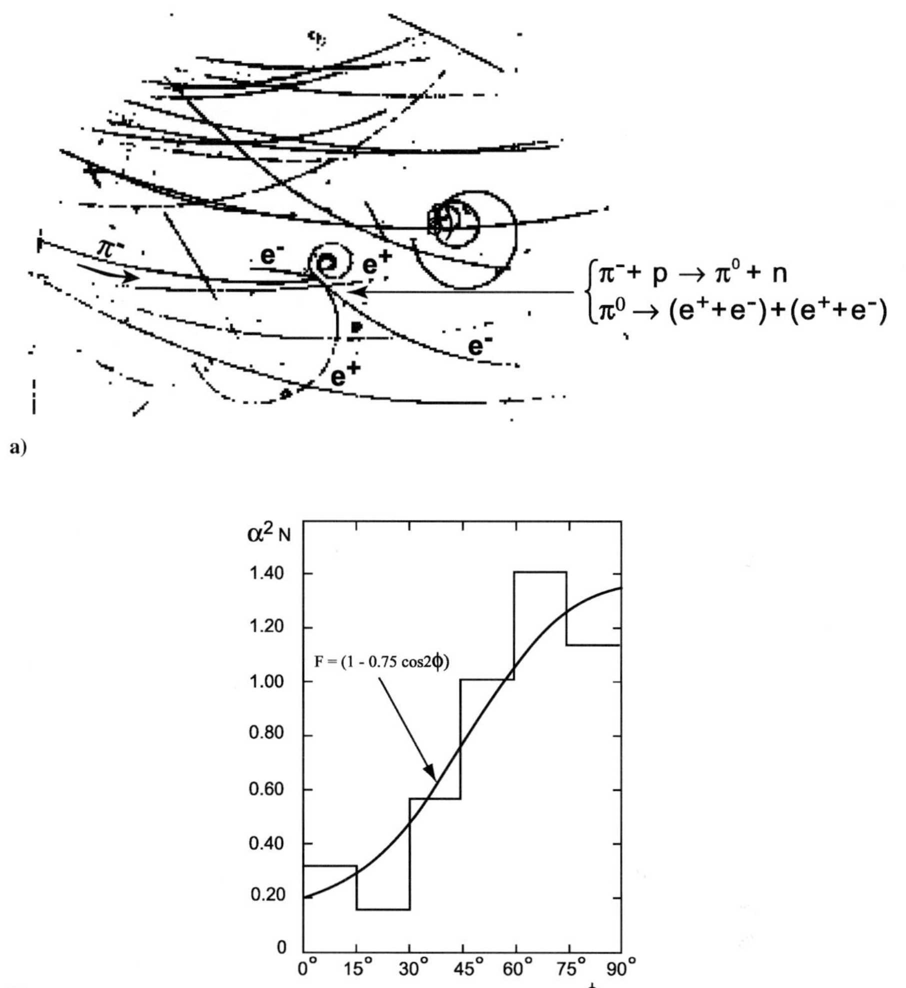

Fig. 4.10. (**a**) Example of the decay $\pi^0 \rightarrow 2e^+ + 2e^-$ in a hydrogen bubble chamber. The bubble chamber is in a magnetic field, which curves the trajectories of the charged particles inversely proportional to their momenta. **b**) Distribution in the angle between the normals to the two decay planes. The solid curve is the expectation for odd parity. For even parity the expectation would have been the same curve but reflected about 45° [34]. The experiment shows that the pion parity is odd

times less likely. Given the close connection to the theory, this branching ratio was of substantial interest. Our first serious search for this decay mode, in 1955, was negative [35], with an upper limit a bit, but not significantly, below 1/8,000. In the experiment the pions were stopped in a thin polyethylene absorber, and the electrons were detected in a scintillation counter telescope after traversing a thickness of absorber. The dominant background was from the electrons produced in the normal pion decay to muon, followed by the muon decay to electron. These background electrons are both less energetic and they occur later with respect to the arrival time of the pion, on average by a factor of 100. The background rejection therefore was based on requiring a minimum range in the absorber and a short time difference between the signals from the pion and electron. The essentially identical experiment, three years later, at the 600 MeV cyclotron at the new European particle physics laboratory CERN, finally gave a positive result [36], and a lower limit of 1/25,000 for the electron/muon branching ratio. At Nevis, we were engaged in a new search and had also seen the decay. We now had the 30 cm hydrogen bubble chamber. Although more tedious, in this technique the muon decay background is essentially eliminated because the muon identifies itself by leaving a visible track, 1 cm long. In addition, the energy resolution of the electron by magnetic curvature was superior to the range technique. At the time of the CERN experiment we had accumulated six electron decay events from 65,000 stopping pions, corresponding to the branching ratio of $1/10,800 \pm 40\%$ [37], in line with the theoretical expectation.

I close the section on a personal, political note. Following the war, which ended rather dramatically with the nuclear bombing of Hiroshima and Nagasaki, there was a substantial increase in the support of physics and, in particular, particle physics, which at the time was lumped together with nuclear physics. In line with long U.S. tradition, the large bulk of physics research was supported under contract with some armed services organization. The Columbia University Nevis laboratory, for instance, was entirely supported by the Office of Naval Research; the Berkeley Radiation Laboratory and the Brookhaven National Laboratory by the Atomic Energy Commission, etc. Although none of the work of which I was aware in these laboratories had any military connection nor was in any way classified, the two AEC laboratories required some low form of clearance; there were fences and guards at the entries. It is a pleasure to remember that this was not true of our navy regime at Nevis, there was no clearance, no fence, nor did the navy ever suggest to us what we should or should not do. The navy was just glad to support us, even with an old navy truck or two when needed[2] and pieces of scrapped battleships for shielding.

Brookhaven National Laboratory was created after the war by the Atomic Energy Commission (AEC) on the location of a former army training camp, Yaphank. It quickly developed into a lively center of particle physics, even before it had, what were during two periods, the world's highest energy accelerators. Particularly during the summers, it would attract a large number of

[2] I am ashamed to admit that sometimes this generous navy help was misused. When Joan, in 1960, decided that it was no longer possible to live with me and, instead, preferred to join a fellow painter in the New York's Greenwich Village, her personal belongings followed in a navy truck.

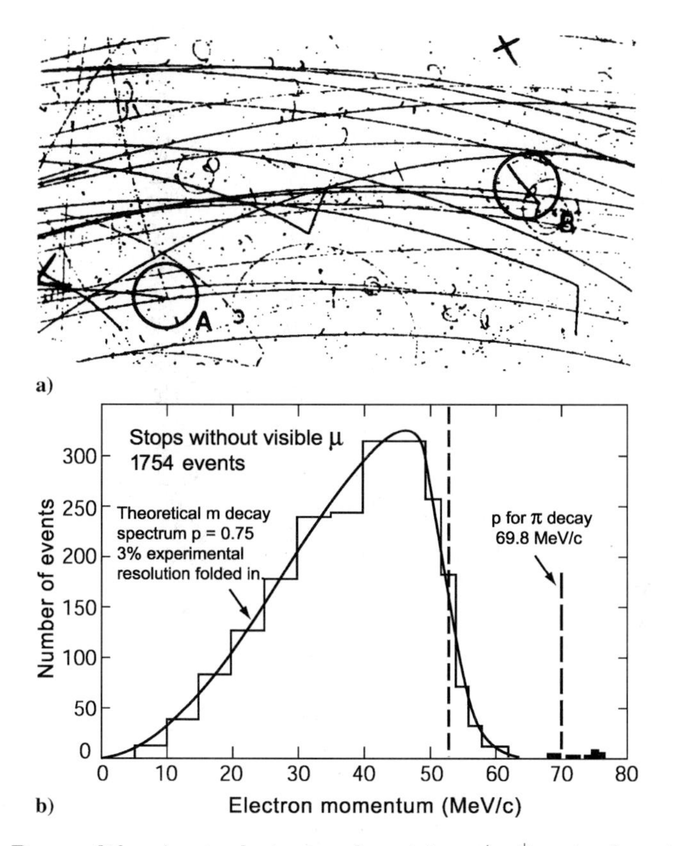

Fig. 4.11. Decay of the pion to electron and neutrino. **a)** π^+s enter from the left and some come to rest in the liquid hydrogen. A marks the decay of a π^+ to a positron; B marks an example of the normal decay chain, $\pi^+ \rightarrow \mu^+ + \nu$, $\mu^+ \rightarrow e^+ + \nu$. **b)** Energy spectrum of the observed positrons. The large peak at lower energy is due to the muon decay electrons. The pion decay electrons are the six events in black near 70 MeV [37]

interesting visitors. I can remember Feynman, Panofsky, Lee, Yang, Serber, and many others who would profit by working together and by having an occasional lobster cookout on the lovely and, at the time, still largely natural nearby Long Island beaches. Brookhaven was only about an hour and a half drive from New York, and so very important to all of us at Nevis, including myself. I was "consultant" to the laboratory. This involved frequent trips during the academic year with transportation paid by the laboratory, as well as summer stay, with housing and summer salary at the charge of BNL.

After a pleasant summer in 1952, it was made known to me – verbally, never in writing – that this association could no longer be continued, since the AEC was unable to renew the clearance. No reason for the security problem was given. I made a trip to Washington, was received by one of the AEC commissioners, but it was a short meeting: "I have looked into your files, there is nothing I can say". It was the Cold War, the country was at the height of its McCarthy political intolerance, and the organization that controlled the bulk of the country's resources for basic physics research did not even consider it necessary to state a reason for the denial of these resources to the researchers. In my work at BNL, no access to any documents of a military nature or marked confidential was involved. Much later, in the aftermath of the 1988 Nobel Prize for an experiment conducted there in 1962, the laboratory invited the three recipients to a celebration. On the occasion the laboratory director gave me a copy of the correspondence of the laboratory with the AEC on this matter. The laboratory had implored the AEC authorities to reconsider, not only because of the importance of openness of basic research, but also because then, in my better years, energetic and experienced, it felt that my presence was valuable in the planning of the experiments for the Cosmotron that was being built and was soon to become operational. The plea was in vain, the response was negative, and the laboratory never received any information on the nature of the difficulty. I consider it important to remember this lack of a sense of accountability of a government that proclaims itself to be a pillar of freedom. Our political intolerance at that time can match the human rights violations that we now deplore in China and elsewhere.

When Bill Nierenberg, fellow professor in the physics department at Columbia, heard of this, he suggested that he propose me as consultant to his research contract with the navy. He was engaged in classified work, I believe for underwater submarine detection. He knew that if the navy had difficulty with granting clearance, the procedure of the naval research office was different from that of the AEC; the person concerned would be given a bill of particulars and the right to an administrative hearing, to which both sides could bring witnesses. On September 10, 1954, I was informed that the navy had difficulties; I was given a statement of three problems, which may be worth recording here to give an idea of the atmosphere of those times:

1. It has been alleged that SUBJECT, while employed at the Radiation Laboratory, MIT, May 1943–June 1945, indicated a sympathetic interest in the Communist movement and had openly stated that he was a Communist.
2. During the period referred to in Item 1, above, it has been alleged that SUBJECT maintained sympathetic association with Floyd Banks and Dirk Struik, reliably reported to have been members of the Communist Party.
3. Investigation disclosed that while residing at 1563 Thousand Oaks Avenue, Berkeley, California, in the period 1949–1950, SUBJECT subscribed to the "Daily Worker," an East Coast Communist newspaper.

This certainly was much more responsible treatment of the problem than the AEC had shown. The list of particulars gave me some clues to the possible sources of part of the information, and at the hearing in New York on March 3, 1955, the key witness had actually been invited by both the board and myself.

The hearing was civilized. The last point was pure invention, at no time had I read, much less subscribed to, the named paper nor could I imagine who the source of this misinformation might have been. Another question which was put during the hearing but had not been stated in the bill of particulars also reflected information in the files that had nothing to do with me. This illustrates what must be a common problem with intelligence information: That it contains much untruth, since it is unchecked. The first statement was also untrue, to the extent that Communist with capital "C" presumably refers to adherence to some communist organization. The interesting question was the middle one. To start with Floyd Banks, he was a physicist, a communist, and also a black (the word was "colored" in those days). Given the widespread racial discrimination, then much harsher than now, and the scarcity of black scientists, I was eager to be as friendly with him as possible. The more interesting connection was that with Struik. Struik was an excellent mathematician. A few years ago, as student, my daughter enjoyed a colloquium given by Struik at Brown University in honor of his 100th birthday. At the time, I was 24, Struik was in his forties and was my professor in a course on Riemannian geometry at MIT. We never had any personal, political, or social contact. Then, during the McCarthy era, he was persecuted as a communist, as were many other intellectuals and artists. He fared better than many, for instance, Frank Oppenheimer, brother of Robert and later architect of the famous San Francisco science Exploratorium, who was dismissed from Minnesota University, or Melba Phillips, dismissed from New York University. MIT did not dismiss Struik, but he was not allowed to teach courses. Later, reading some reflections by himself on his political life, it is clear that he was committed to leftist ideals, but not involved with communist organizations. Where did the intelligence misinformation that I had been a friend of this supposed communist originate? My hunch, which turned out to be correct, pointed to a colleague in our small wartime antenna group at the MIT Rad Lab, formerly a mathematics professor at Queens College, very nice but with no professional pretensions. Politically he was to the Right, I to the Left, and we had many discussions and disagreements on political issues, almost every lunch time, always friendly. Also, otherwise our relationship was friendly; he would often give me a lift to work, and ten years later, he came from Boston to New York to be a witness at the hearing, I am sure out of consideration for me. Why had he volunteered this misinformation? Certainly not with animosity; the reason must be understood within a more complicated psychological frame.

The board found in my favor, and some weeks later Brookhaven also invited me back. But I also like to tell the story of how McCarthy benefited me, although at the expense of difficulties for a friend. McCarthy has helped me greatly in my musical life. In the early fifties, living in Hastings, a well-to-do New York suburb, I came to know the Gobermans living in nearby, even better-to-do Scarsdale. Max was a conductor; at the time (I hope that I remember this correctly), he was in the process of recording the works of Vivaldi with the Vienna Philharmonic. Wife and son were professional cellists. Max was also leftist and, as for many leftist artists, his job fell victim to McCarthyism. So he sold his expensive Scarsdale home and found a more modest old farmhouse in nearby New Jersey (there still was enough money for a riding horse). One day, we came to visit, and there, in the living room, some eight or ten modest wooden chairs were arranged in a semicircle. I asked Max for the reason, and he explained that once a week he invited his nearby farmer friends and gave them lessons on the recorder. So I told him my own frustration with making music. Although as a child I had no instructions in a musical instrument, ten years before in Cambridge, during the war, I had started seriously studying the piano, with lessons even at the splendid New England Conservatory of Music; but after ten years I was nowhere. After diligently practicing things like the Bach two part inventions and some easier Mozart and Haydn sonatas, there was nothing I could play. Max suggested that I try the recorder, which I did. This gave a completely different access to music making. New York being full of musicians, I found friends who would come and play with me. I bought a harpsichord so we could do baroque chamber music. After a while I changed to the transverse flute and, later yet, the harpsichord was traded for a grand piano so we could not only play Baroque, but also classics and romantics. I never was very good on the flute either, but occasionally doing chamber music, thanks to McCarthy, has become part of my life and pleasures.

5

Strange Particles and Bubble Chambers

As mentioned in Sect. 1.4, strange particles were first seen in 1947, in a cloud chamber of Blackett, triggered by hadron showers produced by the interaction of cosmic rays in a sheet of lead above it. One event showed the decay of a neutral particle, called V^0, the other the decay of a negative particle now known as $\sum^-$. Soon after, other strange particles were also seen in nuclear emulsions. Progress in our understanding of these new particles was slow, partly because the experimental possibilities were limited to cosmic ray observations, and partly because the phenomena were so totally outside of what was then known about particles and no one could fit them in. I remember, in 1949 on a bulletin board at the Institute of Advanced Studies, a photomicrograph of a nuclear emulsion event, showing one of these new particles now known as a K^+ meson decaying to three pions. We all saw it, there could be no doubt that something interesting was going on, very different from what was then known, but it was hardly discussed because no one knew what to do with it. Slowly some things became clearer. The decay $V^0 \rightarrow \pi^- + p$ was identified in 1950. This V^0 was then called the Λ°, but could not explain all observed Vs. This was cleared up by Thompson at the 1952 Rochester Conference (these were annual conferences of particle physicists which had begun two years before and which continue to this day) in what was for me a memorable, delicious scene, both because of the clarity and importance of the experimental result, and because of the confrontation of personalities and cultures. C. D. Anderson, discoverer of the first anti-fermion, the positron, was chairman, but, as I remember it, Oppenheimer was in fact dominating. He called on his friends Leighton and then Fretter to give their latest cloud chamber results, but no one was much wiser after that. Some in the audience, clearly better informed than I was, asked to hear from Thompson of the University of Indiana, but Oppenheimer did not know Thompson, and the call was ignored. Finally, there was an undeniable insistence by the audience, and reluctantly the lanky young Midwesterner was called on. His results, also using a cloud chamber, were more precise than those of previous experiments because he had taken particular pains to limit track distortion due to convection. As he started to

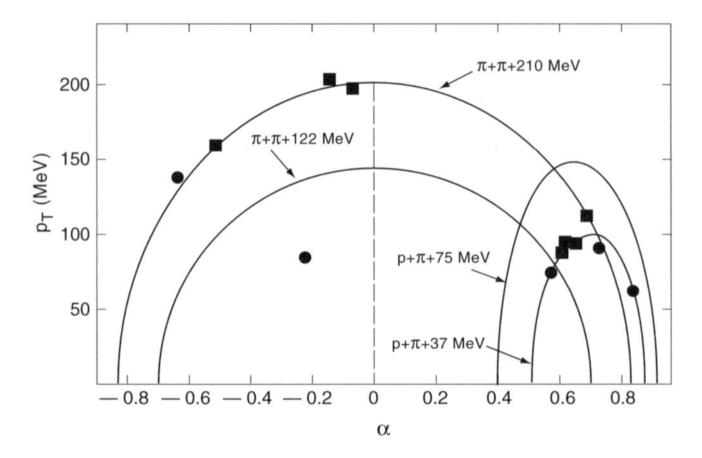

Fig. 5.1. First observation of the K^0, Thompson 1952 [38]. Plot of 12 V^0s. P_T is the transverse momentum relative to the direction of the sum of the momenta of the two particles emitted in the decay, α is the difference of the squares of the longitudinal momenta. The events are seen to lie on two ellipses: one corresponding to the kinematics of the decay of the then known Λ^0, the other corresponding to the new K^0 particle, with mass of about 500 MeV and decaying to two pions

explain some of the experimental niceties, Oppenheimer became impatient, sallied forth from his corner to the speaker to inform him that these details were not of sufficient interest, to get on to the results. Thompson looked at him, then turned to the audience to ask, "Do you want to hear what I have to say?" The audience wanted to hear, and he continued, ignoring the great master's plea. A few minutes later, Oppenheimer again could no longer restrain himself, and so the same scene was repeated for the amusement of the audience. The young man went on, showed a dozen well measured V^0s, and, following a beautiful, original analysis, it was clear that there were two different particles: the known Λ^0, but also another particle which decayed into two pions, then called the θ^0, θ for Thompson, and which is now called K^0 [38]. The result, shown in Fig. 5.1, also gave the value of the new particle's mass to within a few percent.

It began to be clear that the copious production of these particles, indicative of the strong interaction, was at odds with their long lifetimes, indicative of the weak interaction. Pais in 1952 [39] noted that this could be understood by inventing a feature of the strong interaction, a selection rule, which would permit their production but forbid their decay via the strong interaction. He implemented this in a mechanism that required the new heavy particles to be produced in pairs. This was extended some months later by Gell-Mann [40], who ingeniously combined the selection rule with the notion of isotopic spin. It required that the pair of Pais be composed of a "strange" and an "anti-strange" particle.

This was verified a year later in the first accelerator experiment on these new strange particles. The arrival of accelerators of sufficient energy facilitated their study enormously. The Brookhaven Cosmotron had started to operate. It accelerated protons to 3 GeV (billion electron volts), six times that of the highest energy cyclotron, and sufficient to produce the new particles in collisions on nuclei. Ralph Shutt and colleagues had developed a new type of cloud chamber. The V particles produced in cosmic ray showers had been observed in cloud chambers, but these cloud chambers were very inefficient for accelerator experiments because once made sensitive by expanding the gas they would require a minute of relaxation before they could be expanded again. The accelerator cycle however was typically of one second duration. The new "diffusion" cloud chamber, in contrast, was continuously sensitive and it was possible to demonstrate that indeed the production of strange particles is in pairs [41]. One was beginning to understand the physics of these new particles. Figure 5.2 shows the decay in the diffusion chamber of a Λ^0 and K^0, produced together by a pion, $\pi^- + p \rightarrow \Lambda^0 + K^0$. This and several similar events verified the hypothesis of Pais and Gell-Man that the strange particles are produced in pairs.

The experiment also verified Gell-Mann's isospin suggestion. Gell-Mann and Pais then noticed that the neutral kaon should exist in two versions, one strange and the other anti-strange [42], one the antiparticle of the other. In addition to the K^0 which had been seen, there should be another one, with the same mass, but with much longer lifetime and with different decay modes, with opposite symmetry under space inversion. This idea, which seems obvious now, was not obvious at the time. It was not easy for me to understand or accept this proposal when I read it, but a few days later T. D. Lee succeeded in explaining it to me. Once understood, the idea could not be rejected. The experimental confirmation [43] came a year later and marked a big step forward. It was also carried out at the Cosmotron and used what was to my knowledge the largest cloud chamber ever, 1 m in diameter. The large size made it more likely that the long-lived kaon, with a mean free path for decay of the order of 10 m, would decay inside. It had been built at Nevis some years before, but had never found any use. This, to my knowledge, was also the end of the long and glorious career of the cloud chamber in particle physics.

The $K^0 - \bar{K}^0$ system has played and continues to play a singularly important role in our efforts to understand the physics of particles. This special role is connected with the "mixing" of these two particles. For those who may not be familiar, the "mixing" of "states" is a basic feature of quantum mechanics. When the K mesons are created, it is with the help of the strong interaction and they are most usefully characterized by their strangeness, the K^0 with positive, the $\bar{K}^0$ with negative strangeness. However, the particles with definite lifetimes are mixtures of the particles with *definite strangeness*, the short-lived one, $K_s \approx 1/\sqrt{2}(K^0+\bar{K}^0)$, and the long-lived one, $K_L \approx 1/\sqrt{2}(K^0-\bar{K}^0)$. As will be seen later, this is at the root of the experimental demonstration of the violation of the combined symmetry, P = space reflection, simultaneous

Fig. 5.2. Stereo photograph of an event in which a 1.5-Gev π^- produces two neutral V particles in a collision with a proton. Tracks 1a and 2a, believed to be proton and π^-, respectively, are the decay products of a Λ^0. A K^0 is probably seen to decay into π^+ (1b) and π^- (2b). Because of the rather "foggy" quality of this picture tracks 1b, 2a, and 2b have been retouched for better reproduction [41]

with C = charge conjugation, or CP for short. (Charge conjugation is the change of a particle into its antiparticle. This inverts its charge as well as its strangeness.)

In 1953, Donald Glaser invented the bubble chamber [44], a new detection method that went on to dominate particle physics, especially strange particle research, for the next 20 years. He showed that energetic particle trajectories can be made visible by photographing the bubbles which form within a few milliseconds after the particles have traversed a properly superheated liquid. Figure 5.3 shows the first published bubble chamber track, produced by a cosmic ray in a 3-cm-high, 1-cm-diameter glass bulb filled with ether at a temperature of 143 °C and a pressure of 10 atmospheres. In addition to the importance of the invention, this first demonstration was also a most impressive technical feat. The advantage of the bubble over the cloud chamber at accelerators was twofold: The higher density of the liquid proportionally increased the frequency of the interactions produced in it, and it was faster to reactivate, matching the frequency of the accelerator cycle.

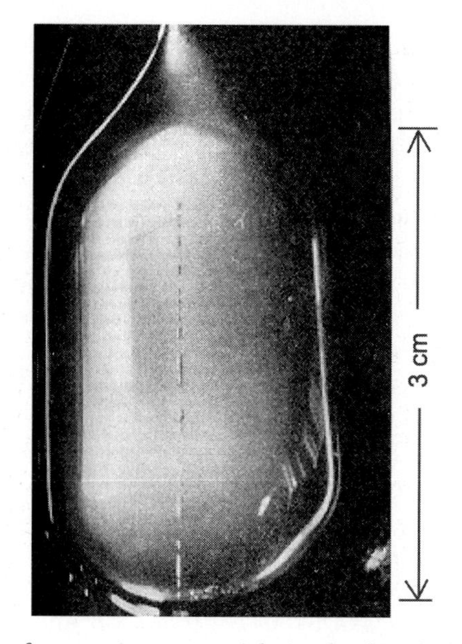

Fig. 5.3. Track of a cosmic ray particle in the first bubble chamber [44]

Within a year, in Berkeley, John Wood [45], in the group of Alvarez, succeeded in producing tracks in liquid hydrogen. The chamber now was a metal cylinder to which glass plates were attached, using indium ribbons as seals. In addition to being a major cryogenic technical achievement, this also demonstrated the crucial fact that for use with accelerators, where the expansion can be timed with respect to the accelerator cycle, the bubble chamber environment need not be as ultra-clean as was the glass vessel of Glaser, which permitted the liquid to survive in its superheated state for relatively long periods.

I do not remember just when we, three graduate students – John Leitner, Nicolas Samios, Melvin Schwartz, and I – began work at Nevis on the design of a practical experimental bubble chamber to study strange particle production at the Cosmotron, perhaps early in 1954. By 1955 we had a 6″ (15 cm) diameter liquid propane chamber [46], Fig. 5.4. This chamber was then used at the Cosmotron in the first experiment using this new technique. The work profited a great deal from a generous collaboration as well as friendship with the inventor, who was working on a similar project at Brookhaven laboratory with his former student, David Rahm [47]. Our main technical contribution at Nevis was the discovery of a rapid-action, three-way gas pressure valve, the "Barksdale" valve. This made it possible to recompress the liquid within milliseconds after the expansion, and so reduce the undesirable thermal effects that result if the pressure remains low for longer times and greater quantities of liquid boil. As work progressed, we were joined by R. Budde, from the newly established European particle laboratory, CERN, who had been sent to learn about the new technique. This developmental work was a great pleasure. The chamber had a serious flaw, however, which we nevertheless accepted in order to get experimental results: The liquid became clouded, and so lost its transparency after a few hours of operation. It was then necessary to empty and refill the chamber, with consequent loss of time.

The experiment [48] used a pion beam of energy 1,300 MeV, only slightly more than the minimum required to produce a strange particle pair. There was no magnetic field, so the particle momenta could not be measured. However, the information from the spatial directions of the observed particles, recorded stereoscopically, sufficed to permit the identification of Λ^0 and K^0 decays, and to distinguish collisions on hydrogen from those on carbon and thereby identify the processes that we wanted to study. Figure 5.5 shows an example of the reaction $\pi^- + p \rightarrow K^+ + \sum^-$. The K^+ is the isotopic partner of the K^0 and decays in this picture to an electron and neutrino. The Σ^- is a barionic strangeness 1 particle like the Λ^0, 5% heavier (25% heavier than the nucleus). The Σ^- was first seen in the original discovery of strange particles in 1947 [5]. The Σ exists in three charge states, Σ^+, Σ^0, Σ^-, and is therefore an isospin triplet. Here it decays to a negative pion and neutron. The picture also gives some idea of one of the very useful features of bubble chambers for strange particle research, in that it permits both the production and the decay vertices to be seen and measured. The lifetimes of most of these particles are of the order of 10^{-10} seconds, and consequently their path lengths are typically some centimeters in the chamber. The several dozen events that were obtained gave the first quantitative measure of the production probabilities and angular distributions for the two reactions, $\pi^- + p \rightarrow K^+ + \sum^-$ and $\pi^- + p \rightarrow K^0 + \Lambda^0$. In retrospect, the most interesting result was a precocious glimpse of parity violation.

One of the major developments of this period was the discovery early in 1957 that the weak interaction violates "parity": It is not symmetric under space reflection, that is, interchanging right and left, or right-handed rotation with left-handed rotation. Before 1954 no particle physicist in his right mind questioned this symmetry. Then, as measurements of their lifetimes and masses became more precise, it became more and more evident that the two particles, the particle then called τ^+, characterized by its decay into three pions, and the K^+ particle, which decays into two pions, must in fact be the same particle. However, the two final states, the three pion state and the two pion state, have opposite reflection symmetry, so at least one of the two decays would have to be in violation of parity symmetry. This was the so-called $\tau - \theta$ puzzle. T. D. Lee and C. N. Yang had the idea that perhaps parity violation is a general feature of the weak interaction and suggested a number of reactions and decays in which parity violation might then be observed [49]. Following one of these suggestions, the violation of parity symmetry was discovered early in 1957 by C. S. Wu and coworkers [50]. In the spring of 1956, while we were analyzing these first bubble chamber pictures for the production of strange particles, some time before the seminal paper of Lee and Yang

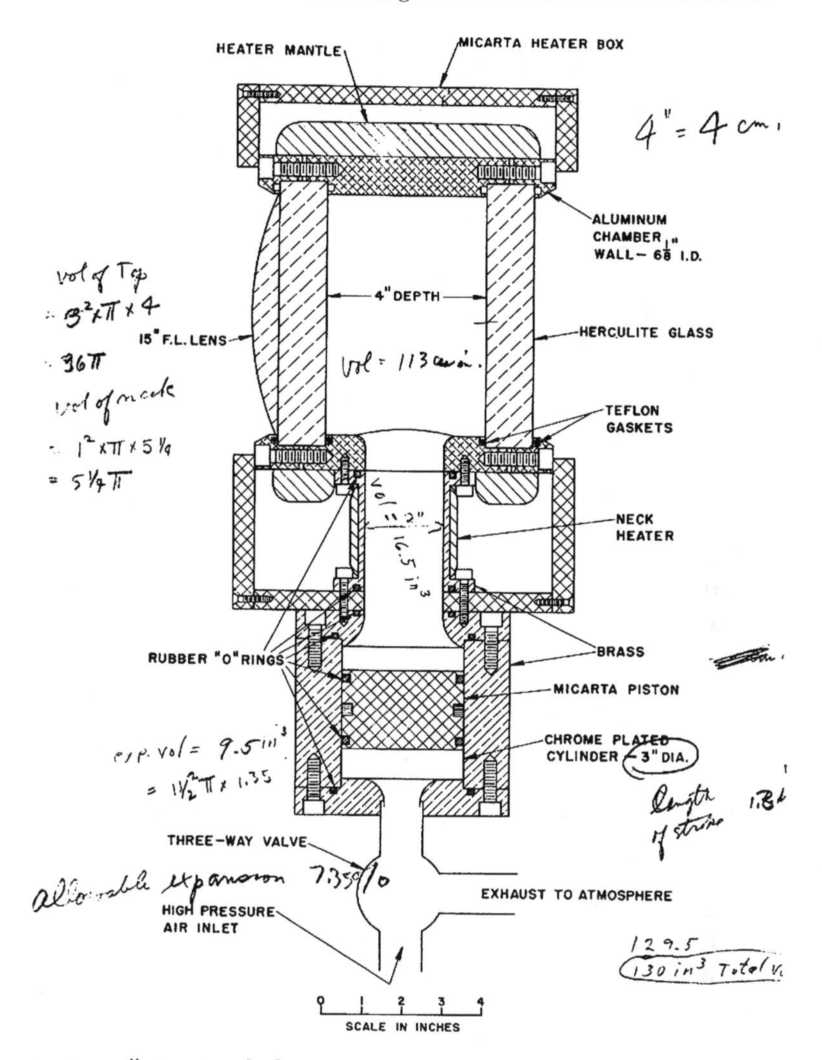

Fig. 5.4. Our 6″ chamber [46]. Notes are contemporary, by Ed Fowler, co-discoverer of the associated production of strange particles, who had studied the drawing carefully before designing and building much larger chambers at Brookhaven together with Ralph Shutt

appeared and the subsequent discovery by Mrs. Wu, we were invited by T. D. Lee to look for a possible parity violating asymmetry in our events, namely, the distribution in the angle (in the Λ^0 center of mass) between the normal to the Λ^0 production plane and the projection of the pion from the Λ decay onto the plane perpendicular to the direction of the Λ^0. An asymmetry about 180° would be a demonstration of parity violation in the decay. The result of the experiment, shown in Fig. 5.6, shows this asymmetry, 15/7. This was

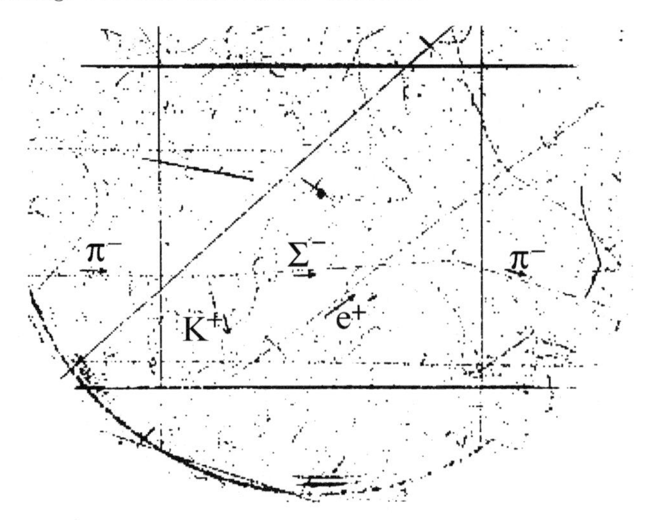

Fig. 5.5. One of the pictures of our first exposure, the associated production by a 1.3 GeV pion of a K^0 and $\sum^-$, on a hydrogen atom of the propane. The $\sum^-$ signs itself by decaying into a slow π^- and an invisible neutron, the K^+ by decaying to an electron and an invisible neutrino

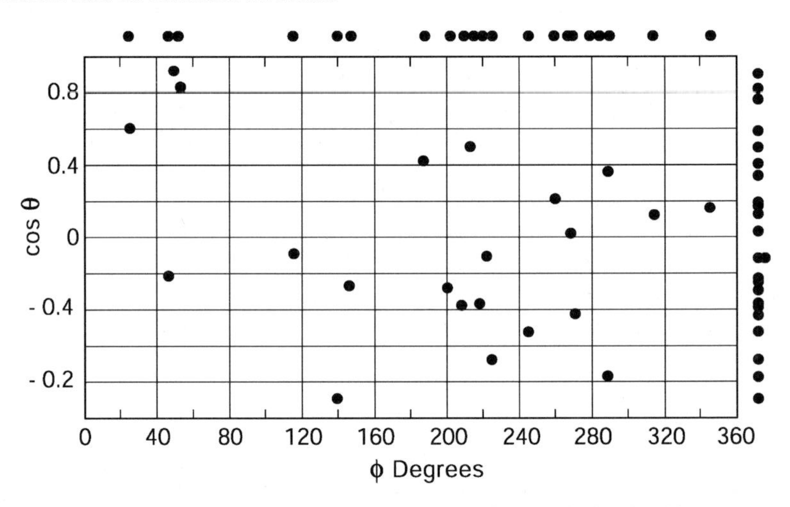

Fig. 5.6. Premature parity violation seen in the first bubble chamber experiment at Brookhaven in 1956 [48]. The asymmetry about $\phi = 180°$ is a manifestation of parity violation

a preview of the discovery of parity violation reported six months later. On its own it was statistically inadequate to permit a conclusion on such a vital question.

The development of bubble chambers went on apace. Within a year the 10″ (25 cm) diameter hydrogen chamber of Alvarez was in operation at the

Bevatron, which was then, with its 5 GeV protons, the world's highest energy accelerator and which had permitted the discovery of the antiproton by Chamberlain, Segre, Wiegand, and Ypsilantis in 1955 [51]. In 1959, this was superceeded by the 72″ (1.8 m) long chamber, the workhorse of the Bevatron for more than a decade and which permitted the discovery of several meson and hyperon resonances. At Brookhaven, the Shutt group made important technical advances. In 1958, its 20″ (50 cm) chamber came into operation, and in 1962 the 80″ (2 m) chamber. This went on to do a great deal of work. Eleven million pictures were taken, and the results included the important discovery, in 1964, of the triply strange hyperon Ω^- [52]. The single serendipitous event, shown in Fig. 5.7, was sufficient to demonstrate the existence of this new particle. It is produced in the chamber by an incoming pion and is seen to decay to a π^- and a Ξ^0 hyperon. The Ξ^0, a doubly strange particle, in turn, decays into a Λ^0 hyperon and π^0; the Λ^0 is seen to decay into proton and π^-, and both photons of the π^0 decay convert in the chamber into electron–positron pairs. This event is the most beautiful demonstration known to me of the power of the bubble chamber in the detection of strange particles. The discovery was particularly important because it dramatically supported the SU(3) flavor[1] symmetry, which had been proposed by Gell-Mann to account for the multiplet structure and mass regularities of the observed strange particles, and which mothered the invention of the quark.

At CERN, a 30 cm hydrogen chamber came into operation in 1960, and the 2 m hydrogen chamber in 1964. The latter became the main CERN tool for the study of resonant and strange particle physics for a decade and kept hundreds of physicists busy and happy. Gargamelle, a very large heavy liquid (freon) chamber, constructed at Ecole Polytechnique in Paris, came to CERN in 1970. It was a vessel, 2 m in diameter, 4 m long, filled with freon at 20 atmospheres pressure, in a conventional magnet producing a field of almost 2 T. As will be discussed later, Gargamelle in 1973 was the tool that permitted the very important discovery of neutral currents. The final bubble chamber was the 20 m^3 Big European Bubble Chamber (BEBC), which could operate with either hydrogen, deuterium, or neon in a superconducting magnet producing 3 T. This splendid machine came into operation in 1975. It was a great and very costly engineering achievement, but it came too late. In the meantime, electronic tracking techniques had advanced and become the more powerful tool.

These bubble chambers took pictures, at the rate of about one per second, on film. Many millions of pictures were produced. These had to be scanned, and the events of interest measured and reconstructed. At first, we used the simple, manual techniques for scanning and measuring inherited from our

[1] Flavor symmetry, the "flavors" referring to the different quarks, is a very important feature of our theory of the strong interaction QCD, which crystallized ~ 1970. Here SU(3) flavor refers to the symmetry underlying the three quarks of lowest mass, called up, down, and strange.

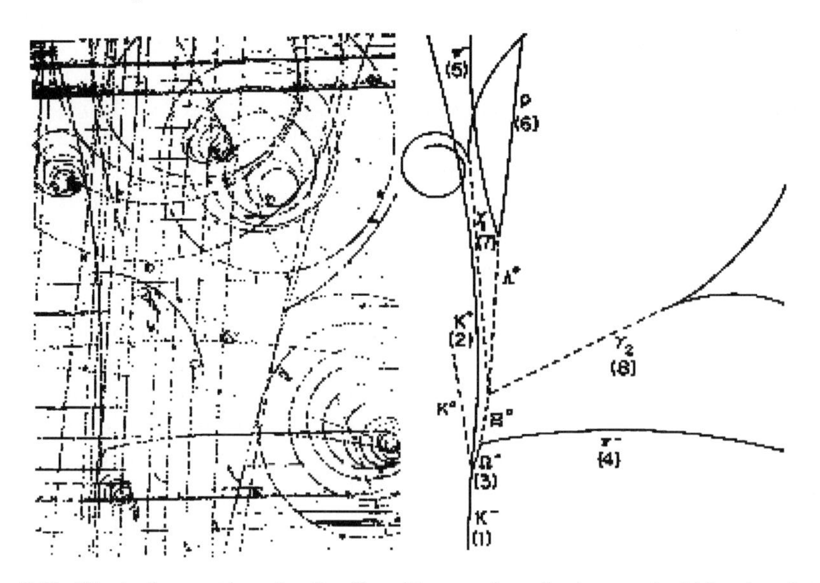

Fig. 5.7. First observation, in the Brookhaven 2 m hydrogen bubble chamber in 1964, of the production and decay of the strangeness 3 hyperon Ω^-. A 5 GeV K^- meson interacts with a proton, producing a K^0 (not seen), a K^+, and the Ω_-: $K^- + p \rightarrow K^0 + K^+ + \Omega^-$. After $\sim$ 1 cm, the Ω^- decays to a π^- and a strangeness 2 Ξ^0 hyperon. The Ξ^0 decays (decay vertex not observed but reconstructed) to a π^0 and Λ^0 :$\Xi^0 \rightarrow \pi^0 + \Lambda^0$. The Λ^0 decays, after 10 cm, into a proton and π^- : $\Lambda \rightarrow p + \pi^-$. Both photons of the π^0 decay, $\pi^0 \rightarrow \gamma + \gamma$, are serendipitously, against the odds, converted into electron–positron pairs: $\gamma + p \rightarrow p + e^+ + e^-$. This permitted the reconstruction of the momenta of all particles in the event, and so even the determination of the Ω^- mass, $m_{\Omega^-} = 1,686 \pm 12$ MeV (compared to m_p = 938 MeV, $m_\Lambda = 1,116$ MeV, and m_{Ξ° = 1,315 MeV)

cloud chamber predecessors: simple projection tables, protractors for the measurement of the angles, templates for the measurement of the track curvatures, and manual computers. But just at that time commercial electronic computers were beginning to appear. We learned to construct digital measuring devices that would automatically punch the track coordinates directly onto IBM punch cards and to write increasingly sophisticated programs that utilized the rapidly evolving power of computers to reconstruct the interesting physical quantities. This was an essential element in the power the technique developed. It was one of the early challenges to the evolving computer industry, and, in turn, the bubble chamber community was able to contribute to the advancement of this technology. Later, the WWW worldwide internet library was invented at CERN.

At Nevis, in 1956, within a few months of the first chamber, we had our first chamber with magnetic field. It was a propane chamber, 12″ (30 cm) in diameter. The volume had increased eightfold, the magnetic field was 1.3 T. One of the technical innovations was the introduction of a third camera so that the field of view was photographed from three angles, rather than two. This was essential to the automatic measurement and reconstruction of tracks parallel to the plane of two of the cameras. In the first exposure we were able to discover the Σ° hyperon, through its decay into a Λ° and a photon [53] (see Fig. 5.8), and so also measure its mass. Together with the previously known Σ^+ and Σ^-, the three formed an isotopic triplet, the first experimental evidence supporting the flavor SU(3) symmetry, and later dramatically confirmed by the Ω^-, as already described. A paper which appeared some months later [54] discusses Λ and K^0 decays on the basis of 528 cases of the production of Λ^0 and K^0. This was a much larger number of decays than had previously been observed. Besides measurements of the lifetimes of these two particles, the non-negligible conversion rates of photons in the propane of the chamber made it possible to see for the first time the neutral decay mode of the Λ^0, $\Lambda^0 \to n + \pi^0$, and of the K^0, $K^0 \to \pi^0 + \pi^0$, and to give first, rough measurements of the branching ratios.

The same magnet as well as optics also served our first hydrogen chamber (see Figs. 5.9 and 5.10), with dimensions similar to the propane chamber. This began operation at the Cosmotron in 1957. The expansion of the liquid was accomplished with the help of a stainless steel bellows, with the associated risk of rupture after many cycles of operation. This would have been an interesting accident involving substantial quantities of hydrogen; nevertheless, I don't remember ever, at the time, trying to understand the likely consequences of such an accident. Soon after, Ralph Shutt at BNL demonstrated that equally effective but safer expansion could be achieved with a piston sealed with Teflon piston rings, and this was the method generally adopted afterwards, also by us. The 12″ hydrogen chamber was used at the Cosmotron, luckily without accident, to continue the study of strange particle production by pions, their decays, and other properties. One of the first results [55] was the demonstration of parity non-conservation in Λ^0 decay, now with about ten times the statistics of the premature experiment of 1956 (see Fig. 5.11). This report combined the results obtained in the 12″ propane and hydrogen chambers, and those obtained in a somewhat smaller propane chamber by my mentor and inventor of the bubble chamber, Don Glaser. Similar results were also obtained by the Berkeley pioneers in the hydrogen bubble chamber development [56] at the Bevatron.

This experiment was followed by a determination of the spins of the Λ and sigma hyperons [57]. It was natural to assume these to be 1/2, the same as those of the proton and neutron, in line with SU(3) flavor symmetry, but this was not known experimentally. In the experiment the spin of 1/2 could be demonstrated, for both the Λ and the Σ, on the basis of the distribution in the angle of production of the hyperon, as well as its decay, following a nontrivial theoretical analysis by Bob Adair [58].

One of the main interests of the bubble chamber community in the early sixties was in the discovery and study of meson and baryon resonances, and the determination of their properties and relationships to each other. Reso-

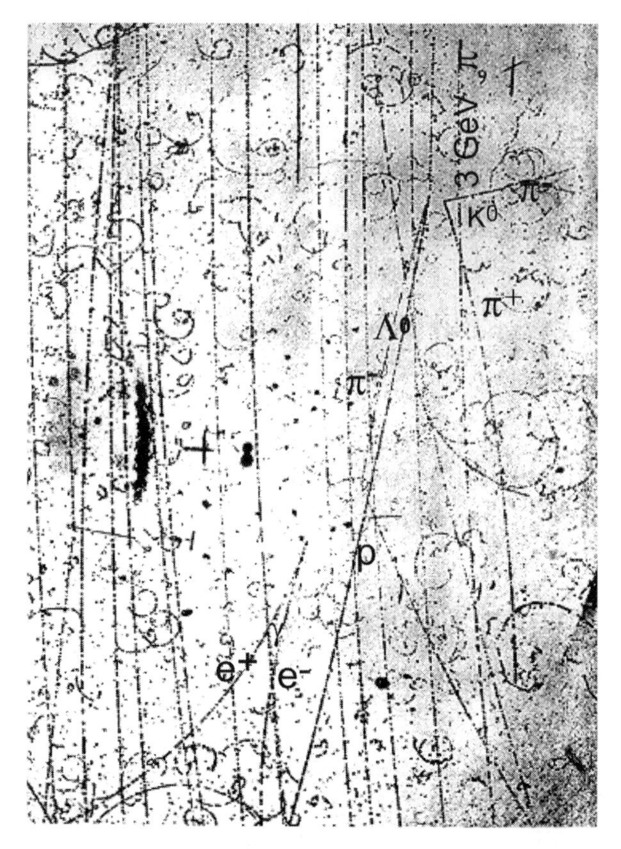

Fig. 5.8. Event seen in the 12″ propane chamber. π^- (1,300 MeV) $+ p \rightarrow K^\circ + \sum^\circ$, followed by the decay $\sum^\circ \rightarrow \Lambda^0 + \gamma$. The photon converts to an electron–positron pair. This is the first observed example of $\sum^\circ$ decay, and permitted a determination of its mass [53]

nances are excited states of hadrons that decay rapidly, through the strong interaction, and therefore have poorly defined total energy or mass,[2] a direct consequence of the Heisenberg uncertainty principle. The resonance "widths," that is, energy spreads of the resonances, are typically of the order of 100 MeV, corresponding to lifetimes of the order of 10^{-23} seconds. In contrast, the muon, pion, kaon, and hyperons that we have encountered up to now decay via the weak interaction, with lifetimes ten or more orders of magnitude longer, and are considered as "stable." The first hadronic resonance to be seen was the spin 3/2, isospin 3/2 baryon resonance in the pion-nucleon system, with mass 1.23 GeV and width 120 MeV, which we encountered in Chap. 4.

[2] The "mass" of a resonance is the "invariant mass" of the decay particles equal to their total energy (divided by the square of the velocity of light) in their rest system, that is, the reference frame in which the sum of their momenta is zero.

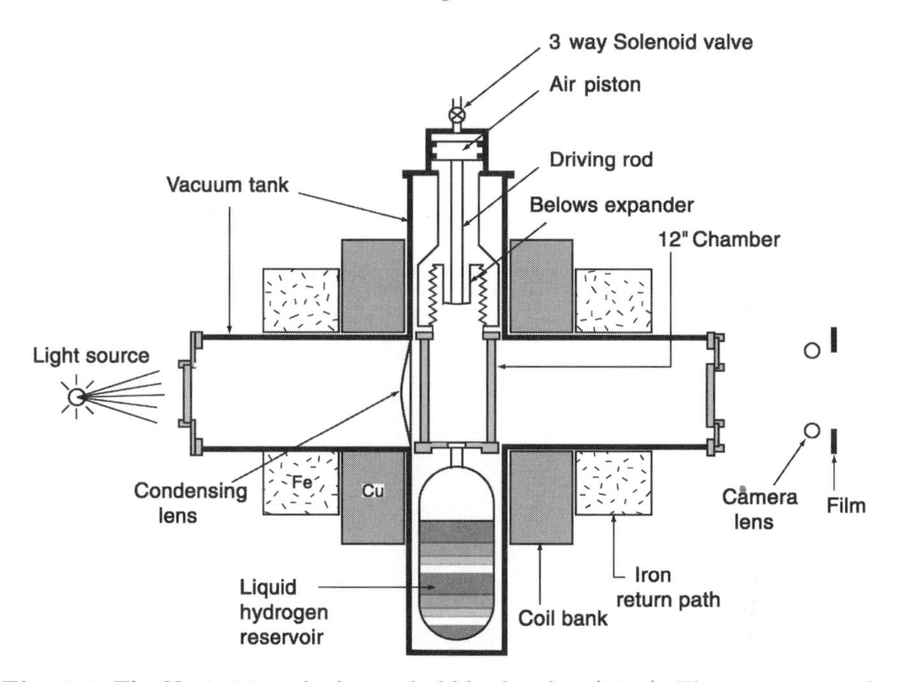

Fig. 5.9. The Nevis 30 cm hydrogen bubble chamber (1956). This gives some idea of expansion mechanism and optics (Nuov. Cim. **X** 471 '58)

The first meson resonance to be seen was the ρ, with a mass of about 750 MeV and a width of 150 MeV, which decays into two pions and exists in three charge states: ρ^+, ρ^0, and ρ^-. It was discovered in 1961 by Erwin et al. [59] at the Cosmotron in the 14″ hydrogen bubble chamber of Adair-Leipuner, using beams of 1.89 GeV π^- mesons and the reactions $\pi^+ + p \to \pi^0 + \pi^+ + p$ and $\pi^- + p \to \pi^- + \pi^+ + n$. The peak in the invariant mass[3] of the two outgoing pions, seen best in the upper part of Fig. 5.12, is the ρ resonance. It could be concluded that the ρ has both spin and isotopic spin equal to 1.

The first hyperon resonance was seen in Berkeley, which then had a 15″ (38 cm) hydrogen chamber in operation at the Bevatron. Berkeley had also pioneered the electrostatic separation of particle species, making use of the difference in velocity of particles previously selected to have the same momentum. In a kaon beam they investigated the reaction $K^- + p \to \Lambda + \pi^+ + \pi^-$ and found a resonance in the $\Lambda - \pi$ system, as seen in Fig. 5.13 [60], with mass 1.38 GeV and width 37 MeV. The Λ–π resonance could be classified as a hyperon multiplet with three charge states, $+, 0$ and $-$, spin 3/2, strangeness -1, and mass 1,385 MeV. It was followed by the discovery of an excited state of the Ξ hyperon, with two charged states, 0 and $-$, spin 3/2, strangeness

[3] It is usual to label resonances by their spin and isotopic spin. In the present case both are 3/2.

Fig. 5.10. Testing the 12″ hydrogen chamber at Nevis

-2, and mass 1,530 MeV, and finally, in 1964, as we have already seen, by the Ω of charge -1, spin 3/2, strangeness -3, and mass 1,672 MeV. Together with the excited state of the nucleon, seen in 1953 by Fermi and Ashkin, with four charged states, $++, +, 0$, and $-$, spin 3/2, strangeness 0, and mass 1,232 MeV, these formed a decouplet, a multiplet of altogether ten states, four of strangeness 0, three of strangeness -1, two of strangeness -2, and one of strangeness -3. Such a symmetry in "flavor" space, between what are now known as the up, the down, and the strange quark, the flavor SU3 symmetry, had been suggested by Gell-Mann some years before, and he had anticipated this decouplet. The discovery of the Ω^-, following the previously discovered resonances, was a clear confirmation of this symmetry, an important step in the evolution of Quantum Chromodynamics (QCD), the present theory of the strong interaction, and is now part of the color SU3 symmetry of QCD.

At Nevis, in the meantime, we constructed two more chambers, again one using propane and one hydrogen. They were 30″ (75 cm) in diameter, substantially larger than their predecessors; but when they came into use, late in 1961, there were already larger chambers in operation. Nevertheless they found some use: Some 12 million pictures were taken in the 30″ hydrogen chamber. In tune with the times, we did some work on the production and

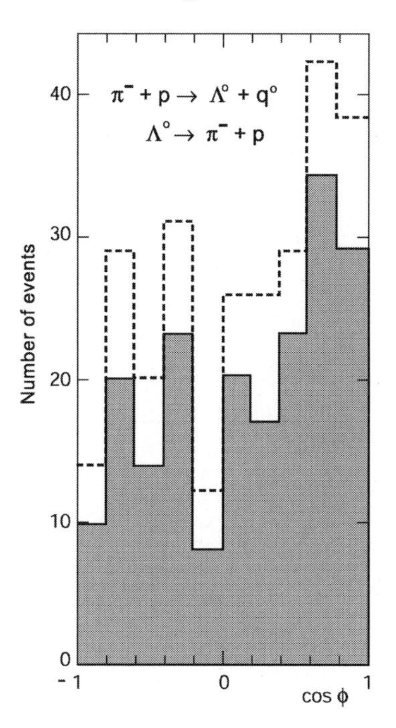

Fig. 5.11. Distribution in the angle between the production and decay angle of the Λ^0. The shaded area represents those events produced in the center-of-mass angular region 30°–150°. The asymmetry about 90° demonstrates the violation of parity in the decay [55]

decay properties of resonances. In one set of measurements, antiprotons of a separated beam were brought to rest in the hydrogen chamber. The antiprotons combine with protons and annihilate, with many different possible final states, of different numbers of pions and kaons, and one could try to exploit these to gain some insights. One rather particular use we made of this exposure was the first determination of the widths of the ω and ϕ resonances. These were among the more interesting meson resonances that had been observed and were distinguished by the fact that their widths, or equivalently, their inverse lifetimes, were too small to be measurable in the usual experiment. In the case of the ω, we were able to select a few events of the type $p + \bar{p} \rightarrow \omega + K^+ + K^-$. Given the masses of the particles involved, the kaons in this process are emitted with such small kinetic energy that they typically come to rest in the chamber. This made it possible to determine their energies very precisely from their ranges, and by energy conservation, the mass of the accompanying ω and its decay width, $\Gamma_\omega = 13 \pm 2$ MeV [61] (see Fig. 5.13). Very similarly, for the ϕ, it was possible to exploit the fact that in the reaction $p + \bar{p} \rightarrow K^+ + K^- + \pi^+ + \pi^-$, where the two kaons are the decay products of the ϕ, the kaons are also emitted with such low energy that they come to

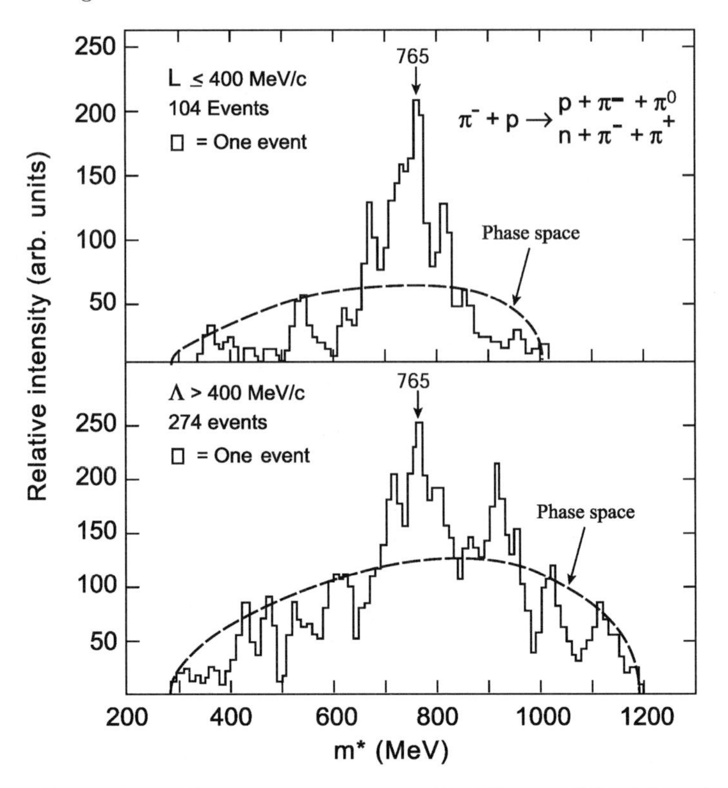

Fig. 5.12. First observed meson resonance, 1961. The combined invariant mass spectrum of the two channels, $\pi^- + \pi^0$ and $\pi^- + \pi^+$ [59]

rest in the chamber. This made it possible to resolve the small width of the ϕ. Γ_ϕ and to find $\Gamma_\phi = 3.1\pm\ 1.0$ MeV [62].

Searching for resonances at random was not my style, and I never looked for nor found a new one. I preferred to focus on something specific, of interest for some particular reason. In one experiment, K^- mesons were stopped in the chamber in order to study the relatively rare leptonic decays of the $\sum$ hyperons resulting from the capture by protons: $K^- + p \to \sum + \pi$. In the same K^- exposure, we could measure the relative parity of the $\sum^\circ$ and Λ° hyperons. Of these 30″ chamber experiments, this was probably the one we valued most highly. This was also the thesis of a doctoral student, Cynthia Alff, who in the meantime had become my wife, after Joan had decided to throw in the sponge, in 1961. In the K^- capture reaction, $K^- + p \to \sum^\circ + \pi^\circ$, normally the $\sum^\circ$ decays to Λ° + photon, but as we saw in Chap. 4, already in the case of π° decay, the photon converts internally into an electron–positron Dalitz pair with probability 1/160, giving rise here to the decay $\sum^\circ \to \Lambda + e^+ + e^-$. The distribution in the momenta of the electron and positron is calculable in quantum electrodynamics. In particular, the expected distribution in the

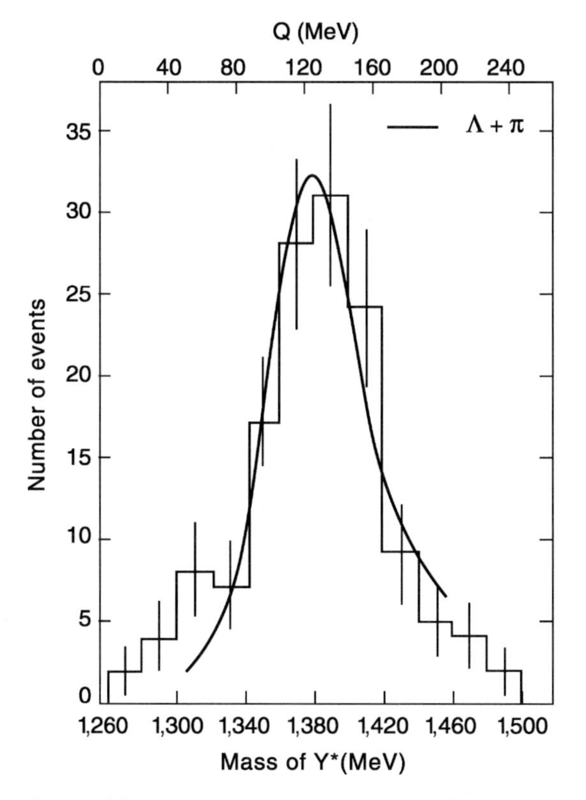

Fig. 5.13. First observed hyperon resonance, 1960. The distribution in the invariant pion-lambda mass shows resonance behavior (*solid curve*) [60]

invariant mass of the pair falls off more rapidly at high mass for even relative parity than in the odd case. In this way it could be seen that the relative $\sum^{\circ} - \Lambda^0$ parity is even [63] (see Fig. 5.15), as expected in the Gell-Mann SU(3) model. An identical experiment was also done at CERN [64].

This was pretty much the end of my bubble chamber adventure. Since my first contact with the field in 1947, particle physics had advanced and changed very much. Cosmic rays were entirely replaced by accelerators, experiments took more time and were carried out by larger groups, more often ten than one or two. Quite a bit had been learned. Four elementary particles had swollen to dozens, the weak interaction had witnessed a wave of clarification in the wake of the discovery of parity violation, and the particle detectors had advanced with the advent of the scintillation counter and the bubble chamber. I had a great deal of pleasure, not only in contributing to our understanding of the physics, but also in the design and even in the mechanical construction of the detectors: counters, liquid hydrogen targets, bubble chambers, even the electronics, where I did not shine particularly.

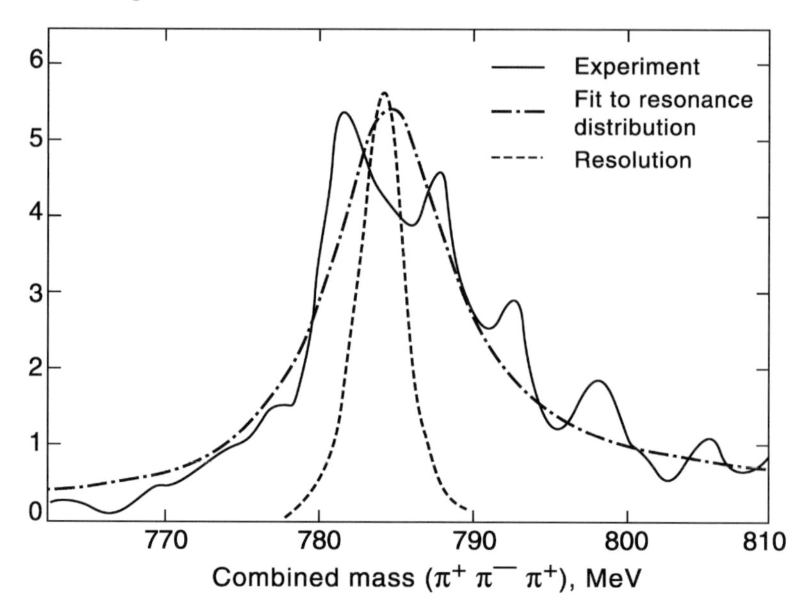

Fig. 5.14. Measurement of the decay width of the ω, which decays into three pions. The ω is produced together with a K^+ and K^- in the annihilation of an antiproton in hydrogen. Each event is represented with its energy resolution (*solid curve*)

The work provided opportunity to renew ties with the old continent. One of the first such occasions was a conference held in Glasgow in 1954, in connection with the electron synchrotron of the university. There I had the pleasure to meet Werner Heisenberg, the inventor of quantum mechanics. During the same year, a summer school had been organized in Varenna, in Villa Monastero, a former monastery with a splendid villa with lovely botanical gardens along lake Como.[4] Fermi was the celebrated star; Bernardini, who had been a colleague at Columbia, was also present. Gianpietro Puppi, the conference organizer, and I became friends, the beginning of a long and fruitful association with Italian colleagues. Figure 5.16 is taken on the steps of the villa. The presence of Fermi was a great local event. The nearby-based motorcycle manufacturer Moto Guzzi took advantage and organized a visit to their factory for the school, which culminated in a dinner in Fermi's honor at a restaurant along the lake shore. As we came to the end of the dinner, there

[4] The villa had been given to the Italian Physical Society by Marco and Rosa De Marchi. I had the fortune to be able to return to Varenna several times in the following years. In the meantime, I had become addicted to mountaineering and had also become friendly with several members of the local Club Alpino Italiano. On the occasion of one of these visits, some of us went to climb the Piz Bernina, and there we spent a night in the Cabana Marco et Rosa, at 3,500 m, one of the highest huts in the alps. It had been given to the Italian alpine club by the same philanthropic couple.

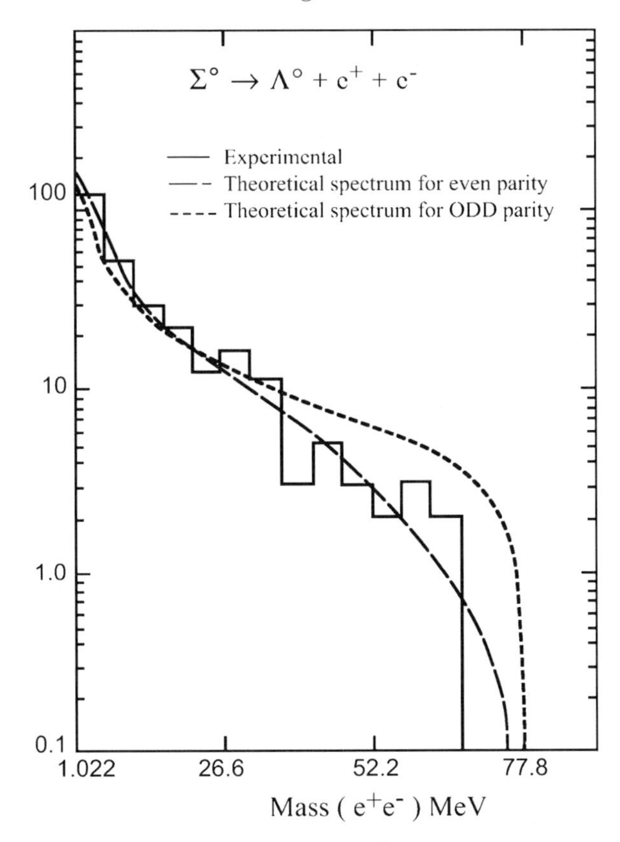

Fig. 5.15. Relative parity of the $\sum^{\circ}$ and Λ^0. Distribution in the mass of the electron–positron pair of the 284 observed decays $\sum^{\circ} \rightarrow \Lambda^0 + e^+ + e^-$

appeared on the lake a string of colored lights, which materialized slowly into gondolas with lanterns and approached. The rowboats landed at our restaurant, and young men and women emerged, dressed in the local costumes of the past century. It was explained to us that they represented characters of a famous novel, *I Promessi Sposi*, written by Manzoni, an author native to the region. The costumed group included musicians, and dancing followed. The evening was a marvelous demonstration of Italian graciousness and hospitality. A photo of the occasion (Fig. 5.17), with Laura and Enrico Fermi in the center, still decorates my bulletin board at work. Varenna is bordered by the Grigna mountains with beautiful chalk cliffs and Fermi enjoyed hiking, so one day a walk was organized. In the snapshot of the group, Fig. 5.18, I can also recognize Eduardo Amaldi, an early collaborator of Fermi's, an important Italian contributor to the physics of particles, and one of CERN's founding fathers. At the time, Fermi began to suffer from the stomach cancer from which he died before the year's end, at the age of 53.

Fig. 5.16. With Fermi and Leprince-Ringuet at the 1954 Varenna school, on Lake Como

Fig. 5.17. Party given at Varenna by the Moto Guzzi motorcycle manufacturer. Laura Fermi is at the table at the right, as seen by the viewer, Enrico Fermi behind her. To the left of Mrs Fermi is Mrs. Amaldi, Eduardo Amaldi is behind Fermi (with permission by Jack Steinberger)

In 1956–57, I took sabbatical from Columbia: One half of the year was spent at the institute in Princeton and the other half in Italy, split between Rome and Bologna. The return to theory at the institute marked my definitive exodus from theory. My colleagues there were doing current algebra and playing with analyticity, both so different from anything I knew as to be inaccessible to me. In Bologna, Puppi was the physics department head, and a group there actively collaborated with us in our bubble chamber work, measuring and analyzing film from the Brookhaven exposures. The excitement of

Fig. 5.18. Hiking in the hills behind Varenna, left to right: Mrs. Puppi, Fermi, the Goldschmidt-Clermont's, Eduardo Amaldi, and I

seeing the remnants in Rome of the Roman and early Christian civilizations was followed by a drive to Sicily, which offered me a first physical contact with our Greek heritage, in the form of the ruins of a Greek temple in the plains of Paestum, near the sea, south of Sorrento, and of Syracusae, the fortified Greek colony in Sicily. The contrast between Greek and Roman aesthetics, as illustrated by the calm beauty of Paestum compared with the business of the Coliseum, made a big impression. The months in Bologna we enjoyed in the midst of our hospitable Bolognese colleagues. Bologna is a beautiful, historic town with a central part that dates from the early Renaissance and before. It is famous for its cooking, especially many varieties of pasta, and one could eat in several bistros most agreeably for 200 lire, a quarter liter of wine included. The best food was at the Puppi's, a household with patriarchal Venetian traditions, which Bianca Puppi had improved with bolognese cooking, served by a white-gloved attendant. The local wine is Lambrusco, slightly bubbly. Bologna is surrounded by lovely hills, the "colline," which were frequent sites for group picnics, with Lambrusco and local pecorino.

The summer of 1957 included a stay at Les Houches, near Chamonix, where Cecile Morette, an old friend from the Institute for Advanced Study, had organized an annual theoretical physics summer school, which became and still is an important part of the French theoretical physics environment. On the occasion, her husband Bryce DeWitt, a well-known theoretician in the field of relativity, introduced me to real mountains. He took several of us, first by cable car to the Aiguille du Midi, then down to the Col de Midi, where we were able to visit what at one time was a laboratory for cosmic ray physics, but which since had been converted into a mountain refuge. We continued down the Vallee Blanche and Mer de Glace: These were my first glaciers. The consequent attraction of the mountains was not unrelated to my subsequent

Fig. 5.19. Landau and Pomeranchuk, Moscow 1956

Fig. 5.20. Left to Right: Lifshitz, Marshak, Smorodinsky, unknown, Gell-Mann, Bakker, and Dyson at the laboratory, Dubna, 1956

frequent visits to CERN, which evolved into a permanent position there in 1968.

An especially interesting trip to Europe was a visit to the Soviet Union, Moscow and Dubna, in 1956. Stalin had died some two years before. It was the first time since the thirties that Soviet physicists could receive a group of Western colleagues. We were a dozen particle physicists, I don't know how we were selected; our hosts were Tamm, Landau, Lifshitz, Alikhanov, Veksler, and Pomeranchuk, Smorodinsky and Feinberg of the Landau school, and others.

Fig. 5.21. Stalin and I

As everyone is aware, Russian physics, especially theoretical physics, was then and is still something of which Russia can be justly proud (I hope fervently that this will still be true in the next decades); but since the thirties, all contact with their Western colleagues had been blocked by Stalin. The meeting was a sign that the Soviet hierarchy was ready to change this isolation. Our Russian hosts were radiant, imagining that this was the beginning of a future in which they would be reasonably free (Figs. 5.19–5.21). I remember the pleasure Landau had in recognizing the genius of Gell-Mann and in discussions with him. I myself, one evening, was in Landau's apartment for dinner. The choice of food was limited, but there were very good strawberries, lots of them. There was no toilet paper in the toilet, but there was a book whose pages substituted for the missing commodity. It was a book on the life of Stalin.

Most of the meeting was in Moscow, but we were also given a tour of the Joint Institute for Nuclear Studies in Dubna, some 150 km northeast of Moscow, in which also the non-Soviet Iron Curtain countries participated. At the time, the jewel of Dubna was a 1 GeV proton synchrotron, built by Veksler, co-inventor with McMillan of the Synchrotron accelerator, but this didn't work very well. Altogether, although Soviet theoretical physics was at the very highest level, experimental particle physics was less impressive, perhaps in part due to the technological lag of the Soviets with respect to the West, and the fact that particle physics had less priority than nuclear weapons.

Fig. 5.22. With Werner Heisenberg at CERN in 1956. In the background, M. Patrascu, a Romanian from Dubna Joint Institute for Nuclear Research

Not long ago, 40 years after the event, a Russian colleague reminded me of an amusing exchange on the occasion of this visit to Dubna. As we entered, behind a very high fence, there was a gang of people in tattered army uniforms, working on some construction. I asked about these people behind the fence, "Were they prisoners of some sort?" The reply: "It is not they, it is we who are inside the fence."

This meeting was a very special occasion. A few years later, in 1959, the annual "Rochester" particle physics meeting was held in Kiev, but this moment of optimism had passed. The opening had closed again, not as tightly as in the Stalin days, but sufficiently to diminish the brighter hopes of our Russian colleagues.

6

Neutrinos I

The neutrino was discovered in the study of nuclear radioactivity. The story begins with the discovery by Chadwick [65] in 1914, using an electron spectrometer, that the electron energy spectrum in β decay is continuous. This raised the question of energy conservation. The fact that the initial energy was not conserved by the visible decay products was verified spectacularly by Ellis and Wooster in 1927 [66], using calorimetry. Clearly, energy, in the form of some new particle, escaped detection in the process. This was underlined by the fact that also angular momentum seemed not to be conserved by one-half unit, as could be deduced from the difference in the angular momenta (spins) of the initial and final nuclei. In 1930, Pauli (Fig. 6.1) dared suggest that if in β decay, in addition to the electron, a neutral, spin one-half particle with very small mass and interaction were emitted, this could solve the problem. He proposed this in his delicious and now celebrated letter to the "Liebe Radioactive Damen und Herren," at a congress in Tübingen. He refers to this "desperate way out," invites the readers to "check and judge" experimentally, while blaming his absence from the conference on a ball in Zurich at which he considered himself indispensable. Only in 1933 did he dare publish the idea. In the meantime, the neutron had been discovered by the same Chadwick [67] who had discovered the electron continuum 16 years before. Before the neutron was discovered, the nucleus could only be imagined as composed of protons and electrons.

The paper of Pauli was followed only a year later by Fermi's quantitative theory, "Versuch einer Theorie der β Strahlen [68]," the very successful theory of the weak interaction, until its replacement in 1973 by the unified electroweak theory. In Fermi's theory, pairs of fermions, electron and neutrino, as well as neutron and proton, in close analogy with electrodynamics, are combined into "currents." In 1949, the muon-neutrino current was added, followed, as they were discovered, by the tau and its neutrino, and the theory became the "universal Fermi interaction." These Fermi currents combine either a charged and a neutral lepton, or an up type and a down type quark so that the charge changes by one unit (in contrast with electrodynamics,

Fig. 6.1. Wolfgang Pauli

in which charged particles are combined with themselves). These are called 'charged' currents. Any two of these currents could be combined to form an interaction, which involves four particles, for instance, the neutron decaying into proton, electron, and neutrino. The predictions could be compared with experiments. For instance, the shape of the β^- decay electron spectrum could be predicted and measured, as well as the lifetimes of different nuclear and particle decay processes. In the forties, discrepancies between the Fermi theory and experiments appeared in the electron spectrum, near its endpoint; but C. S. Wu, in the first of her several very important experimental contributions to our understanding of the weak interaction, was able to show that these discrepancies were the result of experimental inadequacies [69]. The Fermi theory was widely confirmed.

The neutrino became an established particle whose existence could not be doubted. Its experimental verification by means of the observation of some process induced by it, however, had to wait for the arrival of the atomic bomb and nuclear reactors, which, as byproduct, produced strong fluxes of neutrinos. This was accomplished by Reines and Cowan [70] in 1956, at the

Savannah River H-bomb tritium production plant. They observed the reaction $\nu + p \rightarrow n + e^+$, using tanks of liquid scintillator to detect both positron and neutron. The same reaction and technique have become important again in recent years in the search for neutrino oscillations using reactor neutrinos.

Now we know that there is more than one kind of neutrino. The first sign of this was the realization by Feinberg in 1958 [71] that the absence of the decay reaction of the muon to an electron and photon, $\mu \rightarrow e + \gamma$, which had been the subject of sensitive searches, could not be understood if, in the Fermi currents, the neutrino associated with the muon is the same particle as the neutrino associated with the electron. This was followed by the suggestion of Pontecorvo [72] (Fig. 6.2) that one could look for the possible different identity of the muon neutrino by using neutrinos from the decay of pions, $\pi^+ \rightarrow \nu_\mu + \mu^+$, to study the two reactions:

$$\bar{\nu}_\mu + p \rightarrow \mu^+ + n \ ,$$

and

$$\bar{\nu}_\mu + p \rightarrow e^+ + n \ .$$

If the two neutrinos are the *same*, the two reaction rates are expected to be *comparable*; if they are *different*, the second reaction should *not occur*. It was the fourth of Pontecorvo's imaginative and futuristic ideas. It had been preceded in 1946 by the suggestion to look for neutrinos from the sun using the reaction $\nu + \mathrm{CI}^{37} \rightarrow e^- + \mathrm{Ar}^{37}$. A second was the realization in 1947 that the Rome cosmic ray muon capture experiments indicated the universal Fermi interaction. This was followed by yet another highly important suggestion, that neutrinos might oscillate from one species to another and back again. The latter has been just recently experimentally verified, partly as a result of experiments on solar neutrinos. Neutrino oscillations and the associated phenomenon, neutrino masses, are at the forefront of today's particle physics.

Pontecorvo suggested an experiment to check the identity of the neutrino associated with muons, by using the decay of the muons from the decay of stopping positive pions. These could be produced by the Dubna 700 MeV proton synchrotron;[1] however, it was rather clear that the intensities that could be achieved would be insufficient to permit tangible results.

[1] Bruno Pontecorvo, 1913–1993, was one of the most interesting creative colleagues during my long life in particle physics. As already noted, his insights and imaginative ideas were extraordinary in the experimental milieu. From 1931 to 1936 he worked with Fermi in Rome, first as student, then as collaborator. From 1936 to 1939 he was in Joliot-Curie's laboratory in Paris. During the war, he joined the Chalk River Canadian nuclear physics laboratory, where, in addition to the idea of detecting solar neutrinos, he proposed a method of exploring for underground oil deposits using neutrons. He was a communist idealist and emigrated to the Soviet Union in 1950. It is testimony to the respect which he enjoyed that despite the ugliness of the Cold War and McCarthyism, no one, to my knowledge, ever accused him of transmitting nuclear secrets. He was not permitted to traverse the Iron Curtain to visit the West until the eighties. In joining the Soviet Union,

Fig. 6.2. Bruno Pontecorvo (with permission by Jack Steinberger)

Independently, Melvin Schwartz, my former student and associate in several bubble chamber experiments, now a Columbia University faculty member, proposed experiments using neutrinos from pion decay [73], but with a somewhat different physics and experimental perspective. He was stimulated by a question of T. D. Lee, his Columbia colleague: How can we learn about the weak interaction at energies higher than those corresponding to the decays of nuclei, pions, or strange particles (the only manifestations of the weak interaction that were known at the time)? Schwartz, with great insight, realized that, at the higher energies that would become available at the new, nearly completed proton synchrotrons, the CERN PS and the Brookhaven AGS, sufficiently high neutrino flux intensities could be achieved to make the observation of high energy neutrino interactions conceivable. The neutrinos would be the decay products chiefly of pions, but also of kaons, in both cases asso-

he sacrificed his experimental possibilities – in my opinion knowingly, since Soviet experimental facilities were very inferior to those in the West. The search for the muon neutrino identity is a good example. In 1960, this was possible in Brookhaven and at CERN, but not in the Soviet Union. Pontecorvo and I knew each other in 1948 when I was doing my thesis and he was engaged in a similar experiment in Canada, and I much appreciated his generosity toward a junior colleague. In the eighties and nineties, once he was able to travel outside the U.S.S.R., we saw each other several times, in the most friendly way and with great mutual respect; but he never spoke about his reasons for leaving the West in favor of the U.S.S.R. nor the conditions of working there.

ciated with muons. The higher energies of these new accelerators was crucial since:

1. the neutrino interaction probability increases linearly with the neutrino energy,
2. the neutrino beam intensity per unit solid angle increases quadratically with energy, and
3. the detection of the secondaries of the interactions becomes easier and better at higher energy.

The physics interests of such experiments were tabulated in an accompanying paper by Lee and Yang [74] that proved prophetic. Besides the study of Pontecorvo's reactions, the paper listed the search for weak neutral currents (the Fermi currents involved a charge difference in the partners; neutral currents correspond to no charge change, so the proton would be coupled to itself, the neutron to itself, the neutrino to itself, etc.), the question of the point nature of the lepton current, or, equivalently, the possible existence of weakly coupled heavy bosons $W^{+/-}$, etc. These processes became topics of extensive experimentation in the following years, as neutrino beams and detectors became increasingly powerful.

At Nevis, some of us began thinking about how such an experiment might be done. At first, we considered our 30″ heavy liquid-propane bubble chamber as detector, but by the fall of 1960 our focus had shifted to the newly invented spark chamber. In this technique, high voltages are applied across the gas-filled gaps between metal plates, as soon as possible, typically of the order of microseconds, after the particles have passed through, ionizing the gas in the gap along their path. Photographs of the resulting sparks then show the tracks of the particles.

The metal plates would serve as target material in which the neutrinos could interact. In this way much heavier targets than the 200 kg of bubble chamber liquid could be imagined. We ended up with ten tons. I remember constructing at Nevis a first spark chamber model to study the technique. This used the conventional method in which the plates were supported inside a gas containment vessel. Soon this was superseded by a very much simpler construction, an idea of Schwartz, in which the plates were sealed directly to spacers, eliminating the cumbersome external tank. Mel was also the man with the electronics know-how. Producing the necessary high voltages in a sufficiently short time, and without voltage breakdowns, was not trivial. The detector we built consisted of ten one-ton modules, each made of nine 2.5-cm-thick aluminum plates, 110 cm on a side. For the neutrino beam, the AGS 15 GeV[2] protons were allowed to strike a target inside the beam vacuum pipe. The particles so produced were allowed to move freely for 21 m, a distance

[2] The AGS could in fact produce 30 GeV protons, and, as already noted, higher energy is very advantageous from the point of view of reaction rates for these rare processes. Nevertheless, for reasons of inadequate space inside the building, the beam energy was limited to 15 GeV.

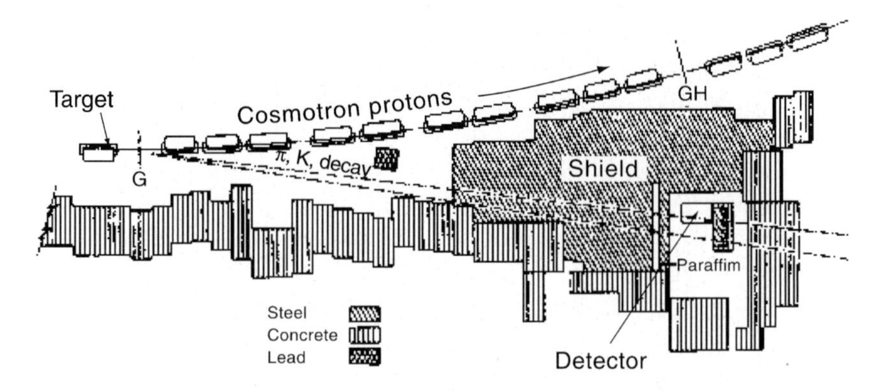

Fig. 6.3. Plan view of AGS neutrino experiment (Danby et al., [75])

sufficient to permit a good fraction of the pions and kaons to decay and so produce the neutrinos, which are emitted predominantly in the forward direction. This was followed by a shield of iron, 13.5 m thick, in which the other particles are absorbed but which the neutrinos traverse unscathed. The detector was behind the shield (Fig. 6.3). Planes of triggering counters were inserted between the chambers and at the end, and a layer of counters was placed above, electronically in anticoincidence, to shield against cosmic rays.

After some weeks of difficulty, due to inadequacies of the shielding, convincing neutrino events were seen [75]. About 60 events were observed in an exposure that took several months. The pictures showed mostly single, penetrating tracks, characteristic of muons rather than hadrons, which interact in the plates with a mean free path of some ten plates, and a somewhat smaller number of events showing also some short nuclear recoil track (Fig. 6.4). Energetic electrons would have produced very different patterns in the spark chamber, since electrons initiate electron–positron showers in the aluminum, with a mean free path of only a few plates. No such events were observed. Muons were typically produced by our neutrinos; electrons were not. This was the most important observation. The neutrinos in this experiment were created in pion and kaon decay, associated with muons: $\pi \rightarrow \mu + \nu$ and $K \rightarrow \mu + \nu$. It could be concluded that neutrinos associated in their production with muons must be different from those associated with electrons, as in β decay. Twenty-six years later, the experiment was recognized with the Nobel Prize for this early contribution to our present understanding of the family structure of fermions (Fig. 6.5).

This first result of experimentation with accelerator-produced neutrinos was quickly verified by experiments at CERN, where a similar proton accelerator was already in operation and a comparable spark chamber detector had been constructed. There were also successful experiments at CERN, using large bubble chambers of more than a cubic meter volume and filled with either propane or freon. Two developments in neutrino beam technology greatly

Fig. 6.4. Tryptich. Lower right: the detector. Top: typical neutrino event. The beam is incident from the left. A neutrino has interacted in the first module. The long, lower track is a muon, recognized by its penetration. The upper track is a hadron. Lower left, the neutrino team at the time, left to right: author, K. Goulianos, J.-M. Gaillard, N. Mistry, G. Danby, a technician whose name I have unfortunately forgotten, L. M. Lederman, and M. Schwartz. This was the invitation to a dinner following the completion of the experiment

increased neutrino fluxes, and so advanced the neutrino experimentation: The extraction of the proton beam from the accelerator made it possible to place the proton target outside the accelerator, and so to use the hadrons, which are emitted directly forward, and which are the most intense; and the invention by Simon Van der Meer (Fig. 6.6) of the "neutrino horn", an achromatic focusing device for the hadron beam.[3]

[3] Simon Van der Meer is a remarkable physicist and engineer, on the one hand, in his imagination and inventiveness, on the other hand, in his competence, both mathematically and technically. His neutrino horn invention was followed by the idea of "stochastic cooling," which permitted the storage of substantial numbers of antiprotons or other rare particles in storage rings, and was crucial in the subsequent discovery of the heavy intermediate bosons in proton-antiproton collisions, for which he was recognized with the Nobel Prize together with Carlo Rubbia in

Fig. 6.5. The neutrino team, 26 years later, in Stockholm (© The Nobel Fondation)

The most important neutrino experiment ever (in my opinion) was the discovery of neutral currents [76] in "Gargamelle" at CERN in 1973. Gargamelle was an immense heavy liquid bubble chamber, 2 m in diameter and 4 m long (Fig. 6.7), in a magnet that produced the very strong magnetic field of 19 kilogauss. It had been designed and constructed under the direction of A. Laguarrigue at the École Polytechnique of Paris and installed in a horn focused beam at the CERN proton synchroton in 1970. In the meantime, in the late 1960s and early 1970s, the unified electroweak gauge theory had evolved into a coherent, beautiful structure. Much earlier, it had been realized that the Fermi theory, although in agreement with all that was experimentally known, could not be adequate, because in higher order calculations[4] it gave infinite results, and even in lowest order, its predictions at very high energy, much higher than those which could be tested experimentally, violated the conser-

1984. The horn was also a very interesting technical challenge in that it required the discharge of a very large current in a short pulse of time through the horn. He succeeded without difficulty at CERN, whereas a similar attempt in Brookhaven failed. In the case of stochastic cooling, working out the mathematics and convincing himself that it really could be done was an extraordinary feat which I am sure that I myself would not have been able to do.

[4] "Higher order" refers to the fact that in particle physics theory, in general it is not known how to calculate the expected consequences of the theory analytically, but only in successive, "perturbative" approximation.

Fig. 6.6. Simon Van der Meer (© CERN)

vation of probability. The new theory was able to avoid these fundamental difficulties; moreover, it unified, that is, it combined, two of the three known particle interactions, the electromagnetic and the weak interactions, into one. This was very impressive for the theoretical community, but the impact on the experimenters was less. On the one hand, we were ignorant (certainly true in my case), and, on the other hand, the predictions of the new theory were identical to those of the old Fermi theory for all that had so far been measured. However, it also proposed radically new phenomena, in particular, the existence of heavy, vector (that is spin 1) bosons with masses of the order of 100 GeV, much heavier than any particles then known (100 times as heavy as the proton), as well as a new weak interaction involving "neutral" fermi currents, which would result in processes such as $\nu_\mu + n \rightarrow \nu_\mu + n$, which had never been seen, in addition to the old "charged" current processes such as $\nu_\mu + n \rightarrow p + \mu^-$, which were contained in the old Fermi theory. At that time, accelerator energies were too low to permit production of the heavy bosons. However, the search for neutral currents was possible. The Gargamelle group let itself be persuaded by its theorist friends to try. In experimental terms, this meant looking for *neutrino events without muons*. This was a nontrivial

Fig. 6.7. The interior of the Gargamelle bubble chamber

experimental challenge. The muons had been useful in the identification of events as being due to neutrinos, and events similar to neutral current events could be simulated by neutrons, a common background. This experimental difficulty is illustrated by the fact that although by that time thousands of neutrino events had been observed, and, as it turns out, the neutral current rates are about one-third of the charged current rates, no previous experiment could be sure to have seen them.

Gargamelle succeeded in finding 102 "muonless" events in a neutrino beam exposure and 64 in an antineutrino beam exposure, and so demonstrated the existence of the neutral current.[5] One of the events is shown in Fig. 6.8. Comparison of the two neutral current rates with their charged current counterparts permitted a first determination of the important free parameters of the new theory, the weak "mixing" angle, or "Weinberg angle," as well as an independent check on the theory. Of all the experimental results that have been obtained in the 45 years of CERN, this one had the greatest impact. Overnight it established the electroweak theory, which is now one of the two cornerstones of our understanding of particles. The other is the formally closely related theory of the strong interaction, QCD, which had come of age in 1973. Together, the two theories are dubbed the "Standard Model." It is sad that the chief architect of this experiment which discovered the neutral

[5] The neutrino is defined as the neutrino emitted in positive pion decay. The particle emitted in negative pion decay is its antiparticle, the antineutrino.

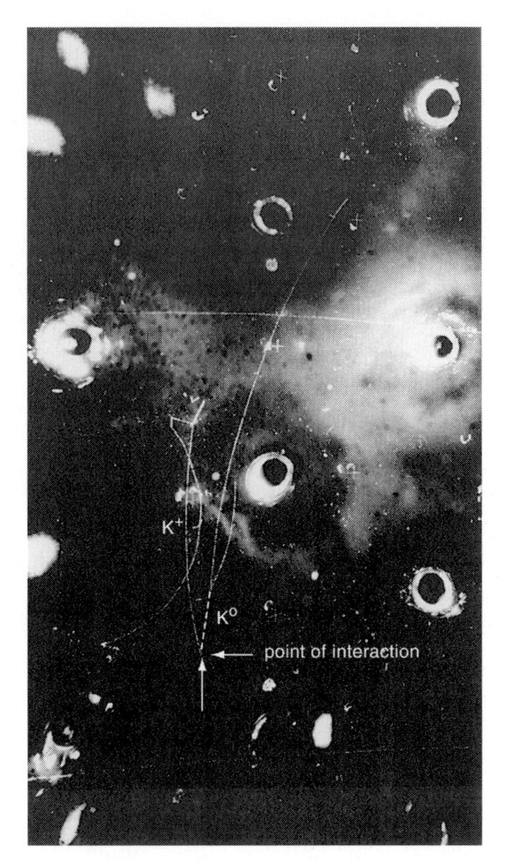

Fig. 6.8. "Muonless" Gargamelle neutrino event in which a K^+ meson and Λ° hyperon are emitted, and nothing else

current, André Laguarrigue, died only a short time later, very prematurely, and so could enjoy the fruits of this great scientific achievement only for a short time.

I permit myself to add a personal note. In retrospect, the experiment is perfectly convincing. At the time, however, I tried very hard to find excuses for not believing the result. I had seen many instances of the charged current interaction, never before any hint of these neutral currents, and I was ignorant of the theory and therefore also of its great theoretical appeal. In addition, I am by nature skeptical and slow in admitting anything new. I made several bets of good bottles of wine with theorist friends, which I was delighted to pay up a few months later with a dinner. The guests included Henry Epstein, Mary Kay and Jean-Mark Gaillard, John Illiopoulos, and Jacques Prentky; the wine included Château Margeau 1959, Château Longueville 1953, and Grands Echezeaux 1969; and Cynthia offered us the best meal ever.

Table 6.1. The three fermion families, their electric charges (in units in which the electron charge is 1), and their masses (in units of 1 GeV = 10^9 electron volt)

Electric	Family 1		Family 2		Family 3	
charge	Name	Mass	Name	Mass	Name	Mass
0	ν_e	$< 3 \times 10^{-9}$	ν_μ	$< 2 \times 10^{-4}$	ν_τ	< 0.018
-1	electron	0.51×10^{-3}	muon	0.106	tau	1.78
$+2/3$	u-quark	$\sim 3 \times 10^{-3}$	c-quark	~ 1.2	t-quark	175
$-1/3$	d-quark	$\sim 6 \times 10^{-3}$	s-quark	~ 0.11	b-quark	~ 4

Weak interaction	Electromagnetic interaction	Strong interaction
$\mu^+ \to \nu_\mu + e^+ + \nu_e$	$e^+ + e^- \to e^+ + e^-$	$u + d \to u + d$

Ten years later, with the help of a new CERN 400 GeV proton accelerator, ingeniously converted into a 270 GeV against 270 GeV proton-antiproton collider, the heavy bosons $W^{+/-}$ and Z^0, were found [77, 78]. This confirmed the other very new prediction of the electroweak theory.

The particles in the Standard Model consist of "families" of four fermions (spin 1/2 particles) each, which are sometimes considered as the "matter" particles, and three types of vector (e.g., spin 1) particles, each of which can be thought of as the vector (in a different sense) of one of the interactions: The massless photon transmits the electromagnetic forces, the heavy $W^{+/-}$ and Z^0 bosons the weak forces, and the eight massless gluons the strong forces. Each family consists of a charged lepton and its neutrino (particles *without* strong interactions) and an up and down type quark (particles *with* strong interactions). For instance, the lightest family consists of the electron neutrino, the electron, the up and the down quark. It is now known (see Chap. 9) that there are just three families (the Standard Model does not give any insight into why just three). All of the $3 \times 4 = 12$ species have been seen. The masses of the particles, within a family and of the different families, are very different. This mass pattern is not understood by the present theory. In Table 6.1 these fermions are listed, together with their electric charges and their masses. In the theory, the typical reaction proceeds via the exchange of a boson between two fermions. For instance, in the "weak" decay of a muon into a muon neutrino, an electron and an electron neutrino, the muon first converts itself into the muon neutrino with the emission of a "virtual" W boson, which then decays to electron and electron neutrino. In the scattering of an electron on a positron by electromagnetic forces, a "virtual" photon is exchanged between the two, and the force that binds the quarks into neutrons and protons, the strong interaction, is due to the exchange of "virtual" gluons between the quarks.

7

CP Violation

When it was shown in 1957 that parity is not conserved, that is, that the laws of physics are not invariant under space reflection, it seemed nevertheless true that the combined operation CP of space reflection P and charge conjugation C was a good symmetry. Charge conjugation is the "reflection" in charge space, that is, the replacement of a particle by its antiparticle, which also reverses the sign of charge and strangeness. As an example, CP symmetry implies that the probability of the decay of a polarized muon to an electron or a positron moving in a particular direction is equal for negative and positive muons polarized in opposite directions. The neutral particles could be classified as either even or odd under this combined reflection. For instance, the short-lived neutral kaon, K_S, was CP positive, the long-lived, K_L, was CP negative. The K_S decays to two pions, a state of even CP symmetry, and the K_L decays to three particles, either three pions or one pion plus a charged lepton and a neutrino, states of odd CP symmetry. However, in 1964 it was seen that this symmetry is also violated. It was observed, at BNL, that there is a small probability that the long-lived neutral kaon, supposedly CP negative, also decays into two pions, a CP positive state, in violation of CP conservation [79]. Figure 7.1a shows the apparatus, a two-armed magnetic spectrometer with spark chamber detectors, to measure the momenta of the two charged particles from the decay of the neutral kaon. The usual decay of the K_L, in contrast with the two-pion decay of the K_S, is into three particles, and this difference was exploited to identify the rare two-pion decay. Because of the momentum carried by the neutral particle, the momentum sum of the two charged decay products in a three-body decay does not in general point in the direction of the beam, and the invariant mass[1] of the two observed charged particles is less than the kaon mass. The evidence for the two-pion

[1] The invariant mass of two or more particles is defined as the square-root of the difference between the squares of the energy sum (including rest energy) and the momentum sum. If the particles are the decay products of some particle, their invariant mass is equal to the mass of the decayed particle.

decay consisted in first selecting those events for which the invariant mass of the pair, assuming both particles to be pions, corresponded, within measurement error, to the kaon mass, and then plotting for these events the angle with respect to the kaon beam of the combined momentum of the two tracks. The peak in the direction of the kaon beam, reproduced in Fig. 7.1b, left no room for doubt that the decay of the long-lived kaon to two pions, though with small branching ratio, does take place, and so demonstrated conclusively that the CP symmetry is violated.

CP violation also plays a major role in our understanding of the Universe, which we know is not symmetric with respect to the exchange of matter and antimatter. Everything we observe is made of baryons, that is, of matter as distinguished from antimatter. Antibaryon stars or galaxies are not observed. This feature of the Universe, essential to the very existence of almost everything we see, including ourselves, must be a consequence of some CP violating process early in the formation of the Universe. It is believed, however, that the mechanism for CP violation in the Big Bang cannot be the same as that responsible for K decay. As yet there is no adequate theoretical understanding of either of these manifestations of CP violation. Until very recently, the only system in the laboratory in which CP violation had been observed was the $K_S - K_L$, or $K - \bar{K}$ system, in which it was discovered. The experimental study of this system came to dominate my work for the next ten years. Now there begins to be evidence that CP violation is manifested also in the decays of other mesons containing b-quarks, and a great effort is going into experiments that study CP violation in B-meson (mesons containing b-quarks) decay.

Within months of the discovery, the phenomenology of CP violation in the neutral $K - \bar{K}$ system was clarified by T. T. Wu and C. N. Yang [80]. K and $\bar{K}$ denote the neutral kaon states with well-defined strangeness, positive and negative, respectively. These states do not have simple CP properties, under CP they transform into one another. The states with simple CP properties are $K_1 = 1/\sqrt{2}(K + \bar{K})$, which transforms into itself under CP, and $K_2 = 1/\sqrt{2}(K - \bar{K})$, which tranforms into its negative. Before the discovery of CP violation, the short-lived and long-lived neutral kaons, the K_S and K_L, were identified with the states K_1 and K_2. In order to understand the CP violation in K_L decay, the K_L, which is dominantly K_2, must additionally contain a small admixture, ϵ, of K_1. ϵ is a complex number. The rate of the CP violating two-pion decay of K_L relative to K_S is proportional to the square of the absolute value $|\epsilon|^2$. As we will see later in detail, this is not exactly true. $|\epsilon|$ was determined to be 0.0023 already by the experiment that discovered CP violation [79]. Similarly, the short-lived kaon, K_S, is dominantly K_1, with a small admixture of K_2. If, even though CP is no longer a good symmetry, the triple inversion, CPT, where T stands for time inversion, does remain a good symmetry, as is an essential property of all modern field theories, then the amount of the admixture for K_S is just the same as that for K_L, that is, also ϵ (i.e., ϵ) so that $K_S = K_1 + \epsilon K_2$.

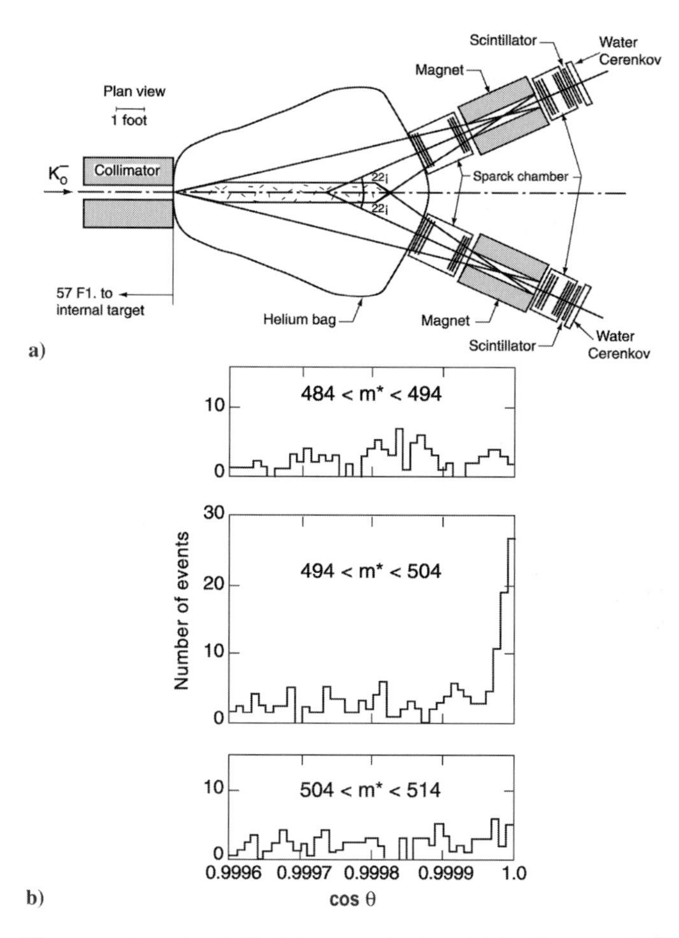

Fig. 7.1. The experiment of Christenson et al., which discovered CP violation. **a**) The detector, consisting of two magnetic spectrometers using spark chamber detectors. **b**) Distribution in the angle of the momentum sum of the two pions with respect to the beam for three domains in their invariant mass: top, invariant mass below the kaon mass; bottom, above the kaon mass; and, center, equal to the kaon mass. The peak in the forward direction, $\cos\theta = 1$, is due to the decay $K_L \to \pi^+ + \pi^-$, and so demonstrates CP violation

While talking with Wu and Yang as they were working on their paper on CP phenomenology in the spring of 1964, I noticed that interesting experiments could be imagined in which the two states, the long-lived and the short-lived neutral kaon, *interfere* in their decay to the same decay mode, which, for instance, could be the two-pion mode observed by Christenson et al. [79], and this could be used to measure the phase of ϵ, a very important quantity in the understanding of CP violation.

Fig. 7.2. Surfing in the Pacific at La Jolla

The experiment had to wait for some months. 1964–1965 was already planned as a sabbatical year: the fall term in La Jolla (San Diego) and the spring term at CERN. The month-long drive to La Jolla, during the summer of 1964 was the delayed honeymoon with Cynthia. We saw the North Dakota Bad Lands, were thrilled by a spectacular sunset at the Grand Canyon's north rim, had an overnight hike down the north face, had a whiff of Pueblo Indian culture at Mesa Verde, and enjoyed the spectacular rock formations of Brice Canyon. In the Sierras, we were met by Willy and Lucette Haldeman for a few days of hiking in the Kings Canyon and Sequoia National Parks. Willy had introduced me to alpine climbing when he was a technician and I an occasional visitor at CERN in the late-1950s; later he came to work with us at Brookhaven in the early 1960s, and now was with the Standard Linear Accelerator at Stanford University. At La Jolla, a new campus of the University of California had just been constructed, only a few hundred meters from the Pacific and the Scripps Oceanographic Institute. The big daily event was the noontime descent to the beach with surfboards. Although my possibilities were exceedingly limited, since surfing requires muscles that the ordinary person doesn't have and which take some time to build up, surfing was addictive, and this was not even discouraged by the advent of winter; it was just necessary to wear a wet suit (Fig. 7.2).

In January 1965, we arrived at CERN for the second half of the sabbatical. From La Jolla I had been in touch with Carlo Rubbia at CERN, future discoverer of the heavy gauge bosons, and we had agreed to join forces in the search for $K_S - K_L$ interference in the decay $K^0 \to \pi^+ + \pi^-$ in an experiment at the 24 GeV proton synchrotron at CERN. We had known each other since Carlo had come to Nevis in 1958 for a postdoctoral year. We proposed an experiment in a neutral beam, with a spark chamber spectrometer triggered by scintillation counters to detect the pions. In many ways similar to the experiment in which Christenson et al. had discovered CP violation [79], but this experiment was to run with and without a "regenerator" in the beam, and the rate of decay to two pions was to be measured as a function of the distance from the regenerator. As in the original Brookhaven experiment, the detector was far from the target at which the kaons had been produced so that the short-lived kaons had died out. The regenerator is a layer of material, in the case of this experiment copper was used. When the long-lived neutral kaons traverse the material, some are scattered or absorbed by the material. This attenuation is different for the K and $\bar{K}$ (strangeness positive and strangeness negative) components of the K_L. As a result, the mixture of K and $\bar{K}$ amplitudes in the transmitted beam is changed; not only is it attenuated, but also it no longer is pure K_L but some admixture of K_S is "regenerated." The detector had also a technical innovation: The magnetic field was chosen so that for the bulk of the decays the pion tracks left the magnet parallel to the beam, and the scintillation counter triggering system was designed to select events with tracks, after the magnet, nearly parallel to the beam (Fig. 7.3a). The advantage of this was twofold: The triggering efficiency was nearly independent of the distance of the decay from the regenerator, and the trigger discriminated against other decay modes.

Since for each event the kaon momentum was measured, it was possible to convert the observed decay distance, following the regenerator, into proper time, that is, time in the rest frame of the kaon, called here τ. The expected time distribution has three terms, two of which decay exponentially with the two respective decay widths,[2] and an oscillating interference term, decaying with the average of the two:

$$|\rho|^2 \, e^{-\Gamma_S \tau} + |\eta|^2 \, e^{-\Gamma_L \tau} + 2|\rho\eta| e^{-\langle\Gamma\rangle\tau} \cos(\varphi_\rho - \varphi_\eta + \Delta m \tau) \ . \qquad (7.1)$$

Here Γ_S, Γ_L, and $\langle\Gamma\rangle$ are the short-lived, long-lived, and average kaon widths, respectively, ρ is the (complex) regeneration amplitude, η is the (complex) amplitude for the long-lived decay to the two pions relative to the short-lived decay. It is very nearly equal to ϵ, as will be discussed later. φ_ρ and φ_η are the phases of the regeneration amplitude and of η, $\Delta m = m_L - m_S$ is the very small difference between the masses of K_L and K_S, about one part in 10^{14} relative to the mass of either, and τ is the time after decay. The in-

[2] The "width" of an unstable particle refers to the uncertainty (distribution) in its mass. It is equal to Planck's constant divided by its lifetime.

terference term has initial phase $\varphi_\rho - \varphi_\eta$ and oscillates with frequency Δm. Data were obtained with two copper regenerators, the first of normal density, 8.9 g/cm^2, the second of effective density, 0.55 g/cm^2, and also without regenerator. The regeneration amplitude is proportional to the density of the regenerator. Figure 7.3b shows the observed time distributions for the three regenerator conditions and Fig. 7.3c the result for the interference term, after subtraction of the two exponential decays and division by its amplitude, $2|\rho\eta|e^{-\langle\Gamma\rangle\tau}$.

The result of the experiment [81] is threefold:

1. The interference is clearly seen. This was the main purpose of the experiment. It demonstrated that the theoretical picture of mixing CP eigenstates in lifetime eigenstates is correct.
2. From the time dependence of the interference oscillation the difference in the two kaon masses was determined with a substantially better precision than was known before.
3. The phase difference $\varphi_\rho - \varphi_\eta$ was measured. As for the latter, φ_η was of great interest, however, since the regeneration phase was not known, φ_η could not be determined either.

We were a bit late with this experiment; before it was completed, the interference had already been shown to exist in an experiment by Fitch and collaborators [82]. The latter experiment was similar in that it regenerated short-lived kaon amplitudes using material of different densities, but simpler in that it did not determine the time evolution of the resultant kaon decays and therefore could not be used to determine either the kaon mass difference or the phase of η.

In the fall of 1966, we returned to Columbia after spending an extra year at CERN following the sabbatical in order to complete the experiment. Initially, my request for a year's extension of leave had been refused by the Columbia physics department, but in the end Columbia decided to accept me back. Until this time, the only process in which CP violation had been observed was that of the original experiment, the decay of the K_L to $\pi^+ + \pi^-$. Some of us decided to look for CP violation in another decay channel of the K_L, the three-particle decay into pion, electron, and neutrino. The effect of CP violation would be seen as a difference between the rates to the two charge channels, $K_L \to \pi^+ + e^- + \bar{\nu}$ and $K_L \to \pi^- + e^+ + \nu$. This CP violating charge asymmetry can be understood as follows: The particles with well-defined strangeness, the K and $\bar{K}$, each decay into leptons of definite charge, the K into electron, the $\bar{K}$ to positron. Therefore, the states of well-defined CP, the K_1 and K_2, each having equal contributions from the two strangeness states, decay into equal numbers of electrons and positrons. However, the admixture of amount ϵ of "wrong" CP in the state K_L upsets this balance so that the small CP violating asymmetry, $\delta = (N_+ - N_-)/(N_+ + N_-) = 2\mathrm{Re}\epsilon$ should be expected.

The new team included Dave Nygren, new postdoc, close collaborator for several years to follow, and future inventor of a very important particle

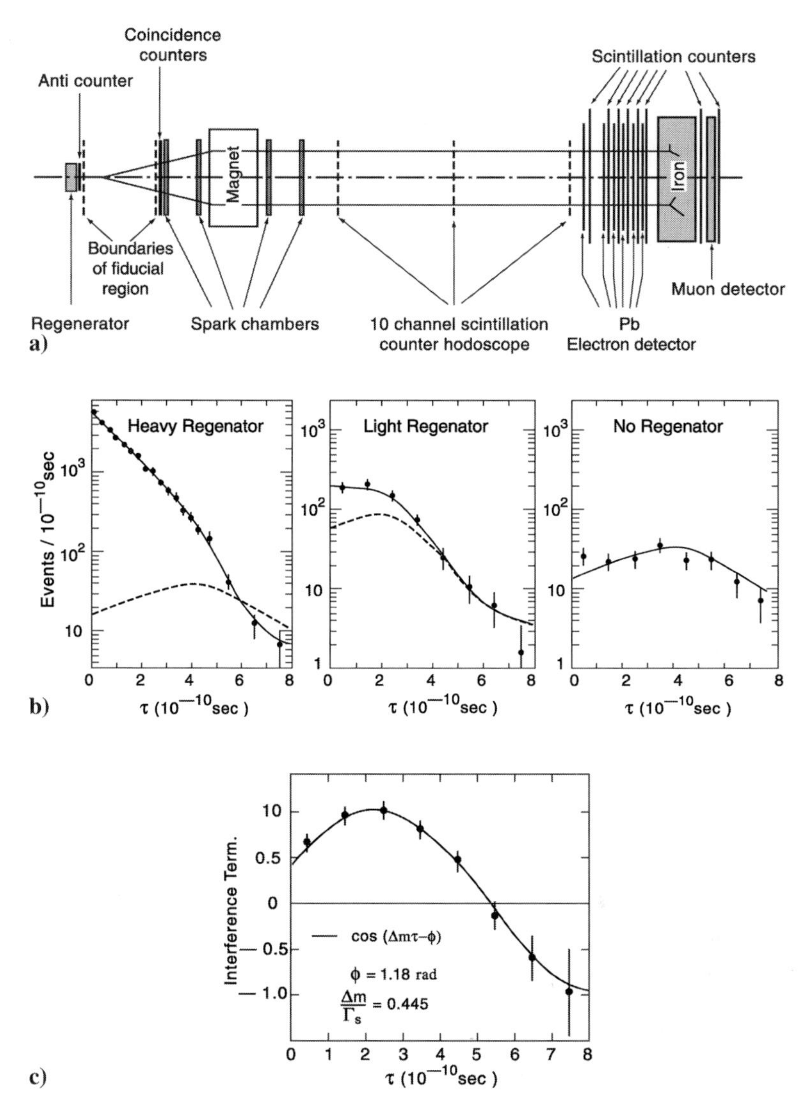

Fig. 7.3. Time dependence of the interference between K_S and K_L in the decay to $\pi^+\pi^-$. **a**) Horizontal plan view of the detector [81]. The triggering counters are narrow, vertical strips; a pair of three in-line counters are required to effect a trigger. The electron and muon detectors following the trigger serve to reject unwanted decay modes. **b**) Time (distance) dependence of the two-pion decays following the regenerator for the two-regenerator densities and for no regenerator. **c**) Time dependence of the interference effect, after subtraction of the pure exponential terms, according to a best fit solution. The phase of this cosine curve is $\phi = \phi_\eta - \phi_\rho$, the difference between the phases of η and ρ; the period is Δm, the $K_L - K_S$ mass difference

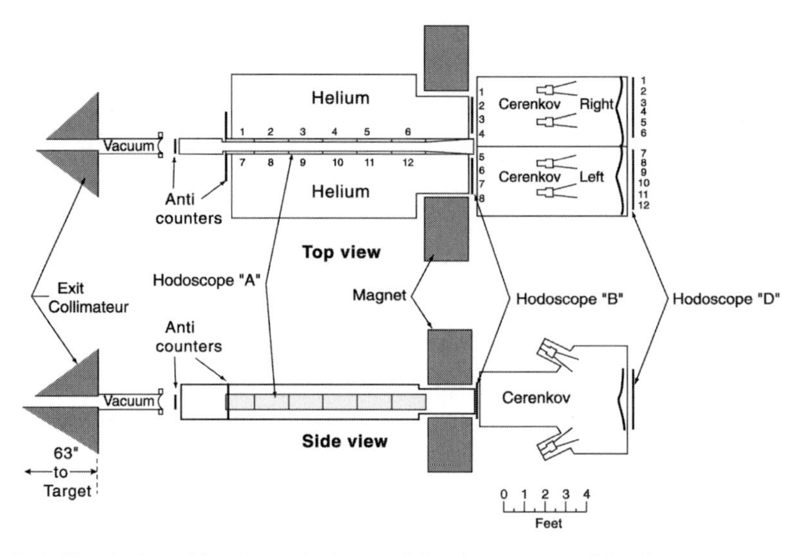

Fig. 7.4. Vertical and horizontal views of the detector used in the measurement of the CP violating charge asymmetry in the decays $K_L \to \pi^{+/-} + e^{-/+} + \nu$

detector, the Time Projection Chamber. On the other hand, Cynthia had left physics to become a molecular biologist at NYU. The beam for the experiment was now at the Brookhaven 30 GeV AGS proton accelerator. The detector is shown in Fig. 7.4 [83]. It consisted of a 4-m-long decay tube lined on opposite sides by a series of thin scintillating plates to detect the two charged decay particles leaving the decay region. This was followed by a magnet to measure the signs of their electric charges and bend them more nearly parallel to the beam. In turn, this was followed by 2-m-long Cherenkov detectors, one on each side, filled with ethylene gas at 1 atm., to detect the light emitted by the highly relativistic electrons or positrons and so distinguish these from the slower pions. At the same time, an experiment was carried out at the Stanford Linear Accelerator, using electrons to generate the kaons, and looking for the similar decay of the kaon, where however the electron is replaced by a muon [84]. Both experiments found an asymmetry, showing CP violation. We found the value $\delta = (N_+ - N_-)/(N_+ + N_-) = (2.24 \pm 0.36) \times 10^{-3}$, Schwartz et al. the value $\delta = (4.05 \pm 1.7) \times 10^{-3}$. The claimed precision of our value was much better than that of the SLAC team, but their result was closer to the present, much more precise value of $\delta = (3.16 \pm 0.12) \times 10^{-3}$.

One of the by-products of the experiment on the charge asymmetry was the possibility of using the same apparatus to measure the regeneration phase. In the experiment in which the interference between long- and short-lived kaons had been observed, only the difference between the regeneration phase and the phase of the CP violating phase, φ_η, could be determined. An independent measurement of the regeneration phase made it possible to get the interesting quantity φ_η itself. The regeneration phase could be obtained from the time dependence of the positron and electron rates, following a slab of material, in our case copper, in front of the detector. It is a consequence of the fact that, as we already noted above, the strange K^0 decays to electron, and the anti-strange $\bar{K}^\circ$ to positron. The regeneration changes the contribution to the kaon amplitude of the two strangeness states, and this is reflected in the difference in the rates of the two charges of the lepton, as a function of time, as the short-lived K_L component dies away. This can be seen in Fig. 7.5, following a regenerator of 30 cm of copper [85]. The observed regeneration phase, $\varphi_\rho = -(29.6 \pm 4.2)^\circ$, when combined with the previous result for the phase difference, $\varphi_\rho - \varphi_\eta = -(80.8 \pm 10)^\circ$, yielded a first determination of the CP violation phase, φ_η, $\varphi_\eta = (51.2 \pm 11)^\circ$.

I will try to give some idea of the important role played by the phase φ_η in the theoretical understanding of CP violation. In the quantum field theories which underlie all understanding of particle physics, there is also the question of the behavior of the equations under *time reversal*, T, that is, the change of the sign of the flow of time, the interchange of past and future. Before the discovery of parity violation in 1957, it was commonly assumed that the theory should be invariant under each of these three symmetries: C, P, and T. It is important here to note that all known quantum field theories have one essential, overriding symmetry, the simultaneous reflection in C, P, and T, that is, a CPT symmetry. The discovery of CP violation raised the fundamental question, is CPT symmetry conserved and therefore is T also violated, or is T conserved and CPT violated? This question can be related to the phase of φ_η with the help of an argument based on the conservation of probability (or unitarity, in the language of field theory) due to Bell et al. [86]. We have seen that the kaon states of positive and negative stangeness, K and $\bar{K}$, can be combined into states of positive and negative CP symmetry, $K_1 = 1/\sqrt{2}(K+\bar{K})$ and $K_2 = 1/\sqrt{2}(K-\bar{K})$. The states with definite lifetimes, the K_S and K_L, but small CP violation are respectively dominantly K_1 and K_2, with *small admixtures* of the other state, $K_S = K_1 + \kappa K_2$, and $K_L = K_2 + \epsilon K_1$, κ and ϵ small. If CPT is a good symmetry, then $\kappa = \epsilon$; if instead T is good, then $\kappa = -\epsilon$. It is now essential to note, firstly, that of all the decay amplitudes of the kaon, the amplitude to the two-pion final state is much bigger than any other, reflected in the fact that the decay probability of the K_S is 500 times greater than that of the K_L, and in addition, as was already known experimentally, that this pion state is largely dominated by a single two-pion state, that of isotopic spin zero. This domination of a single decay amplitude, with the help of the unitarity argument [86], permits the two conclusions: a) $\eta = \epsilon$, and so $\varphi_\eta = \varphi_\epsilon$, and b) $\cos\varphi_\epsilon = \Delta m/\Gamma_S$ for CPT conservation and $\cos\varphi_\epsilon = -\Delta m/\Gamma_S$ for T conservation. Here Δm is the difference in the masses of the long- and short-lived kaons, and Γ_S is the short-lived decay probability. On the basis of the measured values of Δm and Γ_S, the predicted phase angles were 43° for CPT conservation and 137° for T conservation, with an uncertainty of a few degrees. The measured value, 51 ± 11^0, therefore showed T violation and was in agreement with CPT symmetry. These CP violation

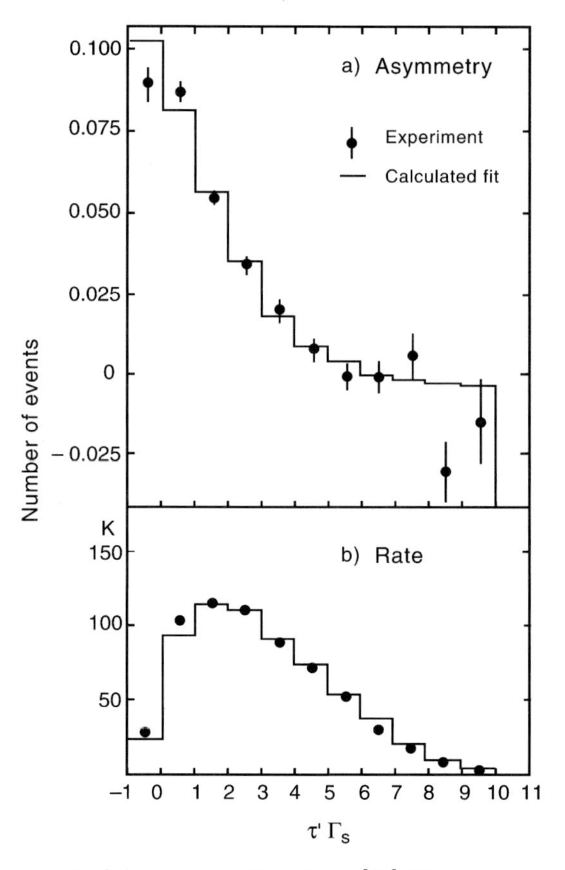

Fig. 7.5. Measurement of the regeneration phase [85]. Charge asymmetry and decay rate as functions of time (distance) in the decay $K^0 \to e^\pm + \pi^\pm + \nu$, following the passage of a K_L beam through a 30 cm copper regenerator

experiments still provide by far the most sensitive checks on the validity of CPT, now with a precision of about 1^0, ten times better.

The short-lived kaon decays with probability 2/3 to π^+ and π^- and 1/3 to $\pi^0 + \pi^0$. Once the CP violating decay $K_L \to \pi^+ + \pi^-$ had been seen, it was very natural and of great interest to look for the decay into two neutral pions, $K_L \to \pi^0 + \pi^0$. Experimentally this posed new challenges, since it necessitates the detection, and measurement of the energies and directions, of the two photons into which each of the two neutral pions decays, and photons are much more difficult to detect than charged pions. The first successful observation, at BNL by Banner, Cronin et al., had to wait until 1968 [87]. Figure 7.6 shows the detector. Of the four photons, only one has both energy and direction measured in a spectrometer, the remaining three photons are detected, and their directions, but not energies, are measured in a spark chamber array. This is sufficient to adequately identify the two-pion decay. The result showed

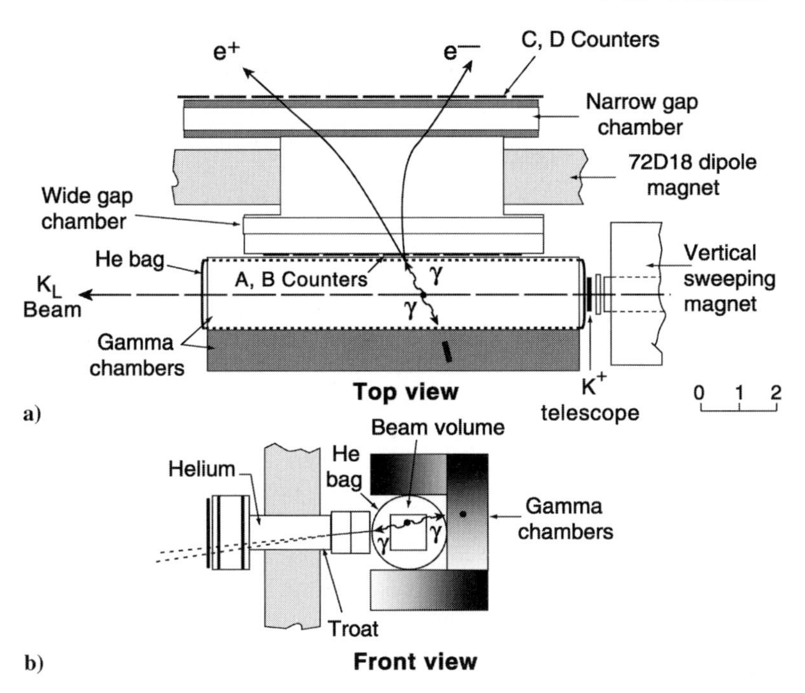

Fig. 7.6. Detector of Banner et al. [87], which permitted the first observation of the decay of the short-lived kaon into two neutral pions. It is shown here as used in the detection of the decay into two photons

that within the uncertainty of the experiment, which was about 14%, the K_L decay into two neutral pions, relative to that of the K_S, $|\eta_{00}|^2$, is the same as the ratio for the decay into two charged pions, $|\eta_{+-}|^2$.

To the extent that both a) the two-pion decay dominates CP violation in K decay and b) that the K^0 two-pion decay is very much dominated by just one of the two isospin channels through which it can proceed, both η_{+-} and η_{00} are expected to be equal to each other, as well as to ϵ. Of course, this implies also that the phase of η_{00} is very close to that of η_{+-}. The first experiment to show this, albeit with limited precision, was in 1970 [88]. A substantially improved determination three years later gave the result $\varphi_{00} - \varphi_{+-} = (7 \pm 18)°$. The most recent, and much more accurate results for φ_{00} [89] are in agreement with φ_{+-}, with uncertainties of less than two degrees.

Six years after the discovery of CP violation, the two CP violation K_L decay amplitudes to two pions, η_{+-} and η_{00}, both their magnitudes and phases, and the charge asymmetries in the decays of K_L into pion, electron, and neutrino as well as into pion, muon and neutrino had been observed and measured, at least crudely. These results confirmed the Wu-Yang phenomenology [80], based on kaon mixing, and enabled a sensitive verification of CPT conservation. A first chapter of our understanding of CP violation had been completed.

In the meantime, Cynthia and I had changed continents, the new for the old. I had changed jobs and Cynthia had changed her field of work. When in 1965 Columbia University could not agree to my request for a year's unpaid leave, I had made some inquiries about other job possibilities, and eventually I also received some offers. Although Columbia graciously took me back anyway, eventually I decided to accept CERN's offer. An important factor in this decision was that CERN was one of the two or three laboratories in the world which offered the facilities for my type of research, and here I could do research "in residence," instead of commuting between laboratory and classroom. The fact that I had a new family and the children of the old were at college and independent probably also played a role in this decision. In retrospect, I gave too little consideration to the fact that this change would separate me from the children in later life, a fact I have come to regret. In my old age, I am useless to them for babysitting their own children.

In 1968, we moved to Geneva, actually into the pastoral French countryside nearby, at the foot of the Jura. After a year's postdoctoral work with the molecular biologists at the university, Cynthia took a job with the World Health Organization. In 1969, I was asked to serve as one of the directors of CERN. The laboratory, with a staff of about 3,000 and an annual budget in present-day dollars of perhaps 500 million, was run by a director general who had full responsibility and authority. CERN was organized into half a dozen departments, each with a director. In turn, the departments were divided into divisions whose day-to-day direction was in the hands of division leaders. As director, I had neither much authority nor a great deal of capability. It was my first managerial challenge, but I had no real interest. I don't remember taking the time to reflect to try to understand how I could best serve the laboratory. I did not give up my own research work and devoted only about half of my time to administration. Since then, watching others, I believe that I have learned that among the most important requirements for good administration are the ability to listen to others and the wish to help them. Neither of these virtues were strongly reflected in my character.

I was also not sufficiently critical of my own understanding of what was going on. The following story is intended to illustrate this. As director I was very influential in the committee which allocated the use of the Proton Synchrotron to the experiments proposed by different physicists. Once there was a conflict between two similar experiments. The first, led by a well-respected and established professor at Heidelberg, had taken data and had been replaced by the second, operated by some young Italians. The allotted time for this was coming to an end, and the first experiment had been scheduled to take over for a second run. However, the young Italians asked for a two-week extension, needed to consolidate a result that they considered interesting. On the other hand, the Heidelberg group let it be known that it would not be possible for it to wait, since they already were scheduled for something else later. To make their point, the Italians had given a presentation to the committee. The committee, on my suggestion, sided with the youngsters. The same afternoon, the Heidelberg professor came into my office to ask, "But Jack, why did you do this?" I replied that the Italian team had shown a graph with a peak on it, indicating an interesting result and that they should have the chance to verify it, and if the professor couldn't wait, too bad for the professor. But "Jack, that graph showed data which had been obtained by *our* group!" He, of course, was right. I had not even bothered to understand the arguments that I had used as the basis of my decision. But in general, in those years, and in contrast to the present, the laboratory had few problems and controversies, so there were not too many occasions for my carelessness to damage it.

The return to CERN in 1968 coincided with the invention by George Charpak of the multiwire proportional chamber [90]. George was an old friend. The house that we had built for our new European life was situated in the fields only a few hundred yards from the Charpak house, by the same artisan who had built his house a few years before and according to a design that was much influenced by Dominique, Mrs. Charpak. The idea of the multiwire chamber was simple enough. Proportional counters had been used since the beginning of the century. They consisted of a wire inside a metal tube, filled with appropriate gas. The wire was operated at a voltage such that the multiplication, in the high electric field near the wire, of the electrons produced by a passing particle was sufficient to make an observable signal, but not enough to produce a discharge, in contrast to Geiger counters, which I had used in my thesis experiment. The voltage of the signal is proportional to the ionization that had been produced by the particle. Since the ionization is a function of the particle's velocity, the pulse height in a proportional chamber can be used to give some measure of this velocity and thereby can help in identifying the particle. Also, in contrast with Geiger counters and spark chambers, there is no down or "dead" recovery time following a signal, the proportional wire is immediately ready for the next signal. The idea of Charpak was to put many wires parallel, in a plane, closely spaced, so that the observation of the wire that has been hit gives a measure of the position of the track passing through the plane.

The proportional wire chamber was an extraordinarily fruitful idea in particle detection. It was quickly followed by the drift chamber [91]: a proportional wire chamber in which one measures also the time it takes for the electrons to drift to the sense wire. The drift speed is of the order of 10^{-8} seconds per millimeter, depending on gas and voltage. The drift time is proportional to the distance of the track from the wire and so makes it possible to interpolate the distance of the track from the wire and so achieve better accuracy, typically 0.1 mm, whereas in the simple proportional chamber, for the typical wire spacing of 2 mm, this is 0.7 mm. Proportional and drift wire chambers, together with the great advances in micro-electronics, revolutionized particle detection in the following years. By the end of the 1970s, the bubble chamber had been completely overtaken.

Not long after Charpak's discovery, it occurred to me that this new idea offered the possibility to measure the CP violating parameters in K^0 decay with substantially better accuracy. It was not obvious that higher accuracy would result in substantial new insight into the question of the origins of CP violation, but for the physicist, in addition to the interest in the underlying physics, also the beauty and technical coherence of an experiment can be an incentive.[3] The proportional chamber would make it possible to achieve much higher rates. It would also replace both the bulk of the trigger counter system as well as the spark chambers in the spectrometer and, in so doing, reduce the amount of material in the path of the particles to achieve higher accuracy in the measurement of each event.

The first order of business was to learn how to make these chambers [92] of the size needed, of the order of 2 m by 1 m, since until that time Charpak had worked only with chambers of the order of 10 cm by 10 cm. A second challenge was that of the electronics. In the design of the chamber we faced two unexpected problems. When we first tried a test chamber with wires about a meter long, it failed; instead of signals there were sparks. The reason turned out to be electrostatic instability. Parallel wires with like voltages repel each other. If the wire is sufficiently short and the mechanical tension is sufficiently high, the situation is nevertheless stable. But above a certain wire length, with a given tension, which is limited by the breaking strength of the wire, the wires push each other apart, they touch the oppositely charged electrode, spark, and vibrate. This was solved by weaving thin nylon threads between the wires every 50 cm or so along their length. The second problem, which surfaced once we could construct a chamber and test it in a particle beam, was the tendency to become dysfunctional after exposure to a certain particle flux. Here we were helped by Charpak, who diagnosed the difficulty as polymer formation and deposition on the wires and who also suggested the cure: the addition to the argon-isobutane gas mixture of a small quantity of some kind of alcohol (dimethoxymethane), which was able to prevent this polymerization.

The electronic requirements represented a considerable challenge at the time, both in sheer quantity and in time resolution. Each wire required its own circuit, and our 5,000 wires needed an order of magnitude more electronics than one had used before. It was designed for us by Bill Sippach, electronics engineer at the Nevis Labs, with impressive elegance and innovation. The 5,000 channels can be compared with several hundred thousand channels in each of the present generation of experiments at the LEP collider at CERN, which was designed 15 years ago, and with several million channels for the

[3] I take the pleasure to repeat here a remark of I. I. Rabi that is relevant to this point. Once, during the time in which I was studying the properties of pions at the Columbia-Nevis cyclotron, I told Rabi of the latest of these experiments. His rather unappreciative response was the question, "But what's the witz?" "Witz" is Yiddish for "cleverness." Rabi's point was clear: To be interesting and fun, it was not enough for him that the physics goal of an experiment be important, he also wanted the experimental technique to show originality.

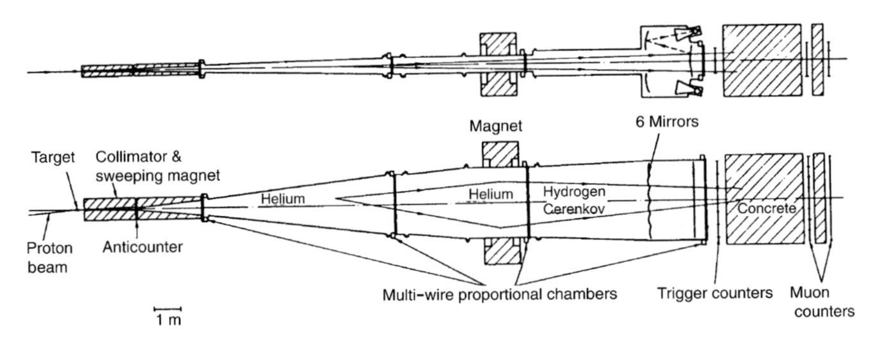

Fig. 7.7. Spectrometer using four multi-wire proportional chambers, each with vertical and horizontal wire planes: top, horizontal view; and bottom, vertical view. Following the magnet, between the third and fourth wire planes, the region is filled with hydrogen gas at one atmosphere and serves as Cherenkov counter to signal the passage of electrons. Similar detectors were built at CERN and at Nevis. The figure shows also the construction of the short neutral beam used at CERN

two large experiments now under construction for the LHC collider, which is expected to operate in 2007.

Two essentially identical spectrometers were built. The team for the CERN detector included Konrad Kleinknecht of the previous CP violation experiment, Heintz Filthuth of bubble chamber days, as well as Heinrich Wahl, just arrived from the German particle physics laboratory at Hamburg, an outstandingly capable colleague with whom I came to enjoy 20-odd years of close collaboration. The Nevis team was based on the group that had worked on the K_{e3} decay, joined by W. C. Carithers and J. H. Christenson, whose Ph.D. thesis had been the CP violation discovery experiment [79]. The Nevis detector was placed into a neutral beam at the BNL AGS accelerator, many meters from the target at which the kaons had been produced, so that the short-lived kaon component had died out. At CERN we constructed a neutral beam as closely as we could to the target struck by the 30 GeV protons of the PS accelerator so that when the kaons arrived in the sensitive region of the detector, enough short-lived kaons survived to permit interference studies without a regenerator. This type of interference experiment had been pioneered by Rubbia and collaborators in 1970 [88]. Figure 7.7 shows the spectrometer as well as the short neutral beam used for the CERN experiment. In addition to the usual muon detector following the spectrometer, electrons could be tagged on the basis of Cherenkov light emitted in a region filled with hydrogen gas at one atmosphere. It was therefore possible to separate the three decay channels, $K^0 \to \pi^+ + \pi^-$, $K^0 \to \pi + e + \nu$, and $K^0 \to \pi + \mu + \nu$, and to compile data in one exposure for all three.

The first result of some interest, obtained at Brookhaven, was a first observation of the relatively rare decay of the long-lived kaon to two muons [93]. On the basis of six observed events, the branching ratio was found to be close to 1×10^{-8}. This observation cleared up a difficulty presented by a previous experiment, performed at the Berkeley Bevatron, that had failed to find the decay and had put an upper limit on the branching ratio at 0.2×10^{-8}, below the "unitarity limit" of 0.6×10^{-8}. The unitarity limit is a theoretical lower limit based only on conservation of probability and a previous measurement of the K_L decay rate to two photons. Violation of the law of probability conservation had been embarrassing. Our result was above this lower limit, and so removed this difficulty. Participation in this experiment turned out to be my last contribution to physics on the new continent.

The main results concerning CP violation from the CERN detector were measurements of the time dependence of the charge asymmetry in the two leptonic channels [94, 95] (Fig. 7.8), the time dependence in the decay into two charged pions [96] (Fig. 7.9), and a determination of the $K_S - K_L$ mass difference using regenerators [98]. The underlying physics for all of these measurements was well known before. However, it was possible to obtain substantial improvements in precision. This can be appreciated if one compares Fig. 7.3 with Fig. 7.9, or Fig. 7.5 with Fig. 7.8. The new values of the parameters relevant to CP violation were:

K_S lifetime	$\tau_S = (0.894 \pm 0.005) \times 10^{-10}$ s,
$K_L - K_S$ mass difference	$\Delta m = (0.5335 \pm 0.0024) \times 10^{10}$ s^{-1},
Absolute value of CP v. amplitude	$\lvert\eta_{+-}\rvert = (2.30 \pm 0.035) \times 10^{-3}$,
Phase of K_L CP v. amplitude [99]	$\varphi_{+-} = (45.9 \pm 1.6)^0$,
Real part of CP v. mixing parameter	$\mathrm{Re}\epsilon = (1.67 \pm 0.08) \times 10^{-3}$.

The pleasure in these experiments was partly in the new precision which could be achieved, but chiefly, I believe, in the exploration of a beautiful, more powerful experimental technique, and the opportunity of contributing to its development. They also supported our understanding of the phenomenology of CP violation, sharpened the confirmation of CPT conservation, but did not open a new window on our understanding of the theoretical basis of CP violation. The same beam and apparatus also served for several other experiments not bearing on CP violation. Since such short-lived neutral beams contain Λ hyperons in amounts comparable to those of the short-lived kaons, it was possible to do some experiments with Λs. In fact, the first result, in 1972, was the measure of the total nuclear interaction cross sections of the Λ hyperon [99]. This was the first experiment using a beam of hyperons. Another experiment making use of the Λ "impurity" of our kaon beam measured the transition probability of the Λ^0 to the $\sum^0$ hyperon in the coulomb fields of lead and nickel nuclei. The $\sum^0$ is a strange baryon, part of an isospin triplet containing also positive and negative partners, with mass about 80 MeV heavier than the Λ, and which decays into a Λ^0 and a photon in a time too short to measure. The coulomb-produced $\sum^0$s can be distinguished from those produced in nuclear processes by their peaked production in the forward direction. The uranium or nickel converters were inserted into the neutral beam at the front of the detector's acceptance region, and a photon detector, consisting of 84

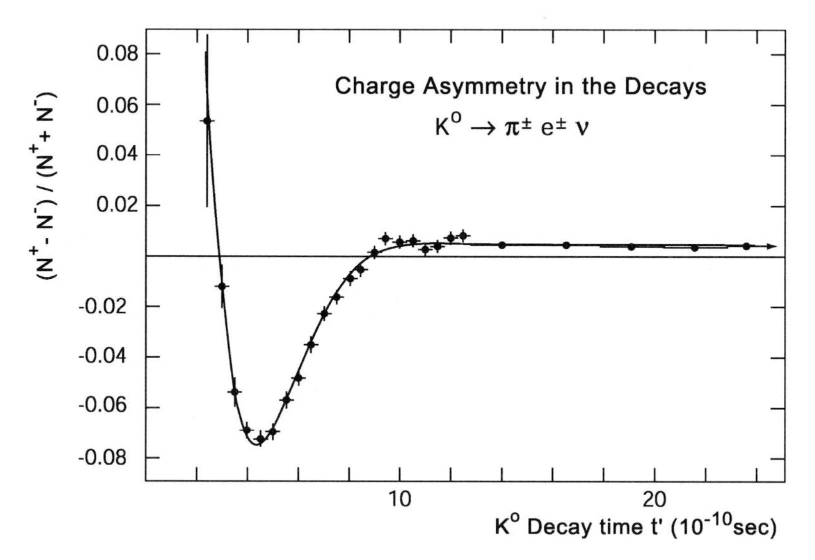

Fig. 7.8. Charge asymmetry in the decay $K \to \pi + e + \nu$ as a function of time in the kaon rest frame, in a kaon beam close to a target in which preferentially K rather than $\bar{K}$ are produced. The oscillation at short time permits the determination of the $K_L - K_S$ mass difference, and the asymptotic value at large times is the CP violating K_L asymmetry [96]

lead glass blocks covering the area of the spectrometer, was inserted between the spectrometer and the muon detector. On the basis of some 100 events above a background of 150 with the uranium target, and 40 events above a background of 60 in nickel, the transition magnetic moment could be determined. Since this moment is directly responsible for the $\sum^0$ decay, the result could be given as the $\sum^0$ lifetime, $(0.58 \pm 0.13) \times 10^{-19}$ s [100].

Another experiment not connected with CP violation but possible in this apparatus was a first "determination" (in quotation marks because not very accurate, less than two standard deviations) of the electromagnetic interaction of the K^0 [101]. Even though it has neither electric charge nor magnetic moment, the kaon interacts with the electromagnetic field since it is composed of charged objects (quarks). The resulting scattering amplitude of the kaon on an electron is equal but of opposite sign for K and $\bar{K}$ and so gives rise to a coherent regeneration amplitude that adds to the nuclear amplitude. It can be sorted out in principle, because the dominant nuclear amplitude can be measured independently and subtracted. What remains, an effect of not much more than 1%, can be assigned to the electromagnetic scattering of the kaon on the electrons and can be characterized by an average of the square of the radius of the electric charge density, normalized to the charge of the electron. It was found to be $\langle R^2 \rangle = (0.08 \pm 0.05)$ fm^2 (it will be remembered that the fermi, fm, is a unit of distance characteristic of hadron sizes and is equal to 10^{-13} cm). The fact that the result was much smaller than the square of the

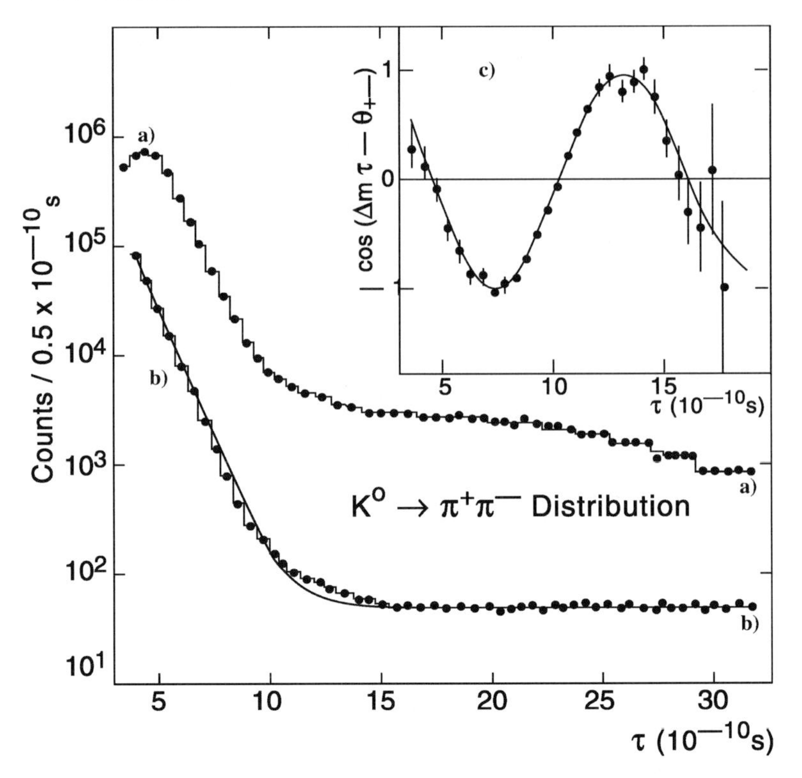

Fig. 7.9. Time distribution of the two-pion decay in the same beam as Fig. 7.7. **a**) is the observed distribution (histogram) and the distribution fitted to the two exponential terms plus an interference term (*dots*). **b**) shows events corrected for detection efficiency and the fitted distributions with (*dots*) and without (*solid line*) the interference term. The insert **c**) shows the interference term as extracted from the data (*dots*) and fitted (*line*) [96]

fermi probably reflects the fact that the quark charges are much less than 1. This experiment, at the time, was at the limit of what was then possible.

Recently, at CERN, the same parameters that we measured with so much pleasure and precision in the early 1970s were measured even more precisely, and in a quite different environment, in an experiment called CPLEAR. The experiment took advantage of the reaction in which antiprotons are stopped in a small vessel filled with hydrogen gas. A small fraction of the subsequent annihilation proceeds via the two charge conjugate channels $\overline{p} + p \to K^+ + \overline{K}^0 + \pi^-$ and $\overline{p} + p \to K^- + K^0 + \pi^+$. In the experiment both the charged kaon and the pion are identified so that it is known that in one case a K is born, in the other a $\overline{K}$. Each evolves with time and will show the CP violating interference effect, but the interference is of opposite sign for the two. By taking their difference, the interference term is directly isolated. The

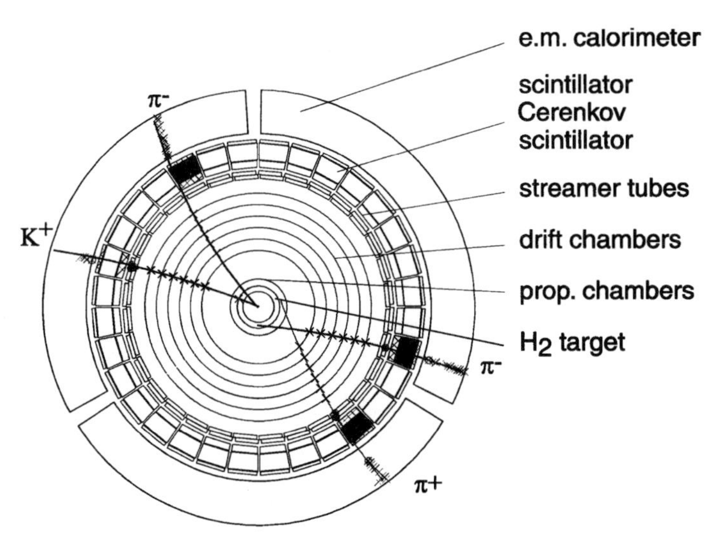

Fig. 7.10. A typical event observed in CPLEAR, which shows a negative kaon and a positive pion from the annihilation vertex (center of apparatus), and a K^0 decaying to a pion pair on the opposite side, some 20 cm distant

detector consists of a cylindrical multi-wire drift chamber in a magnetic field, to measure the momenta of the charged kaon, the pion, and the pions from the decay of the neutral kaon, as well as Cherenkov light detectors and a time of flight system to distinguish pions and kaons. Figure 7.10 shows a typical event. The time dependencies observed for the two strangeness states are shown in Fig. 7.11 [102]. The result is not only beautiful, but it also gives more precise determinations of both $|\eta_{+-}|$ and φ_{+-} : $|\eta_{+-}| = (2.264 \pm 0.023)$ and $\varphi_{+-} = (43.19 \pm 0.53)^0$.

To come back to earlier days, after the series of experiments in the first half of the 1970s, it took a few years for the idea to mature that due to a very interesting experimental question concerning CP violation in K decay there was a possible difference between the two amplitudes η_{+-} and η_{00}. This would answer a basic question: Can all CP violating phenomena in K decay be understood in terms of the $K_1 - K_2$ mixing in K_S and K_L alone, that is, in terms of the parameter ϵ only, or are there, in addition, CP violating interactions that contribute "directly" to the two-pion decay of the K_L. The phenomenology of the possible effect of such direct CP violation had already been analyzed in the original paper of Wu and Yang [81].

If mixing is the only manifestation of CP violation in K^0 decay, then η_{+-} and η_{00} are both equal to ϵ, and so equal to one another. A difference between the two would be a manifestation of direct CP violation. In the search for a possible difference, the square of the absolute value of the ratio of these two amplitudes, $|\eta_{+-}/\eta_{00}|^2$, has been studied in the last decade and a half in a series of increasingly sophisticated experiments. The quantity $|\eta_{+-}/\eta_{00}|^2$

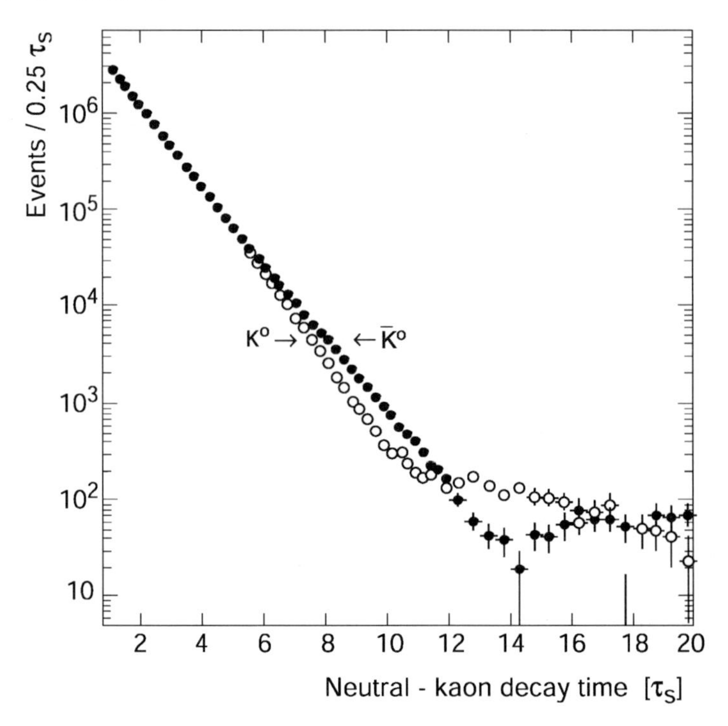

Fig. 7.11. Time distribution, measured in units of the K_S lifetime, of K^0 and $\bar{K}^0$ decays to charged pions, as observed in CPLEAR, showing the CP violating interference, of opposite sign for the two

is just the double ratio: the ratio of the decay rates of the long-lived kaon to charged pions divided by short-lived kaon to charged pions, and this ratio divided by the similar one for decay to neutral pions, $R = |\eta_{+-}/\eta_{00}|^2 = [(L \rightarrow +-)/(S \rightarrow +-)]/[(L \rightarrow 00)/(S \rightarrow 00)]$. It is usual to characterize this by the quantity called $\mathrm{Re}(\epsilon'/\epsilon) = 1/6(|\eta_{+-}/\eta_{00}|^2 - 1)$.

The measurement of ϵ'/ϵ is of prime interest for the theoretical understanding of CP violation. When parity violation was discovered, in 1957, it was very quickly clear how this must be incorporated into the Fermi theory of the weak interaction, and it became a key element of the present unified electroweak theory. The situation is less beautiful for CP violation. In the Standard Model, in combining the strong and electroweak interactions, as well as quarks and leptons, it is necessary to introduce a somewhat awkward "rotation" of the quark states, or what amounts to the same, the Kobayashi-Maskawa "mixing matrix," named after its inventors. This involves the ad hoc introduction of three numbers (mixing angles), which describe the combinations in which the quarks combine in the weak interaction to form the heavy, charged, weak boson, the $W^{+/-}$. It is possible to introduce a fourth ad hoc number into the K-M matrix, this one imaginary, and so produce CP violating effects involving quarks such as the CP violation in K^0 decay. This theory also predicts "direct" CP violation, and so also ϵ'/ϵ. The confrontation of this theoretical prediction with experiment is therefore of great interest.

The determination of the double ratio requires the measurement of four decay rates. Of these, the K_L to two π^0s poses the greatest experimental challenge. In the early 1980s, when these double ratio experiments began, this had been measured with a precision of about 10%; improvement by a factor of more than ten was necessary to achieve a meaningful result. Two groups, one at Fermilab and one at CERN, launched themselves into the pursuit of the double ratio. Both experiments were aided by the fact that, in the meantime, new, more energetic accelerators, at both labs, had increased the available kaon beam energies by a factor of more than ten. This facilitated the detection, especially of the difficult neutral decay. Also, the steady improvement in the capacity of electronics and computing was an essential element in tackling the much higher event rates that were needed to achieve the desired accuracy. At CERN, Heinrich Wahl and I, prompted by a remark of Alvaro de Rujula, decided to think seriously about how we might measure the double ratio. We were joined by Italo Manelli, Konrad Kleinknecht, and younger colleagues. Soon we were some 50 physicists from five laboratories, compared with the ten from two labs of the previous experiment. This reflected the increasing complexity of particle physics.

We arrived at a design in which the four measurements were done in two steps, one in a long-lived kaon beam, the other in a beam produced close to the detector, so that the two pion channel was dominated by the short-lived decay. The detector measured both charged and neutral decay modes simultaneously, so that in determining the double ratio, the kaon beam intensities cancelled and therefore did not require measurement. The chief challenges for the detector were first that of making adequate rates possible, a thousand times those of the earlier experiments and, second, to achieve higher purity, that is, better rejection of unwanted decay channels, in particular in the $K_L \rightarrow 2\pi^0$ channel. It should be stated that the rate problem was not that of producing beams of sufficient intensity (more than adequate intensity was available). Rather, it was the electronic problem of the speed of data acquisition. The accelerator was the CERN SPS 400 GeV proton synchrotron, in operation since 1976. Typical kaon beam energies were between 50 and 120 GeV, corresponding to mean free decay paths of the short-lived kaon of 3 to 7 m. To help achieve the desired rates for the long-lived kaons, the decay region was taken to be quite long, 80 m. The detector (see Fig. 7.12) consisted of the usual elements: the decay region, now requiring a vacuum, since, otherwise, given its length, there would be too much interaction of the kaons in the gas; wire chambers to measure the directions of the charged decay tracks; a photon calorimeter; a hadron calorimeter; and, finally, a plane of counters to tag muons, the only particles (except neutrinos) that traverse the calorimeters. The transverse dimensions were 2.4 m $\times$ 2.4 m. In addition, outside of the volume traversed by the particles to be measured, counters in anti-coincidence were strategically placed to help reject unwanted background events.

Since this is the first time we encounter calorimeters, it is necessary to say a word about them. They play increasingly important roles in particle detection as the energies get higher. In a calorimeter one, first of all, requires that the incident particle deposits all of its energy, by interacting with the material of the calorimeter, producing secondaries, which in turn interact and so on, producing a shower of secondaries. The length and lateral dimensions of the calorimeter must be sufficient to contain this shower. These showers are either electromagnetic, generated by and consisting of photons, electrons and positrons, or hadronic, generated by and consisting mainly of hadrons. The electromagnetic shower is propagated by the electromagnetic, and the hadronic shower by the hadronic interaction of the particles with the material of the calorimeter; consequently, they require different construction. The typical experiment has both, the electromagnetic calorimeter preceding the hadronic, since it is considerably thinner. In the electromagnetic shower, photons create electron–positron pairs in the coulomb field of a nucleus, which in turn radiate more photons, and so on. The mean free path for electromagnetic interaction of photons and electrons is roughly inversely proportional to the atomic number, in lead it is 6 g/cm^2. The required thickness, determined by the shower development, is typically about 25 of these mean free paths, or 13 cm of lead, for electromagnetic calorimeters. For hadron calorimeters a very common material is iron. The mean free path for hadronic collisions in iron is about 120 g/cm^2, or 15 cm, and about 1 m longitudinally and perhaps 25 cm transversely are typical requirements for the containment of hadronic showers. Many techniques are used for the measurement of the energy that is contained in the shower. By far the most common is the sampling of the ionization produced by the particles, by means of scintillation counter planes or wire chambers. More rarely, measurement of the ionization produced in liquefied inert gas, such as argon, has been used.

Our detector was exceedingly simple in the track detection and hadronic calorimeter elements. The former consisted only of two wire chambers 25 m apart, each containing four wire planes: horizontal, vertical, and $\pm 45°$. This was indeed simpler than any previous detector for this decay, all of which had used magnetic spectrometers to measure, in addition to their directions, the momenta of the two particles. However, the precise knowledge of the two directions was sufficient to determine the necessary information for the event:

a) to distinguish the two-pion decay from the background, since the plane of the two-body decay, by momentum conservation, must contain the beam direction, whereas this is not the case for the three-body decays that constitute the background;
b) to reconstruct the momentum of the parent kaon; and
c) to reconstruct the point of decay.

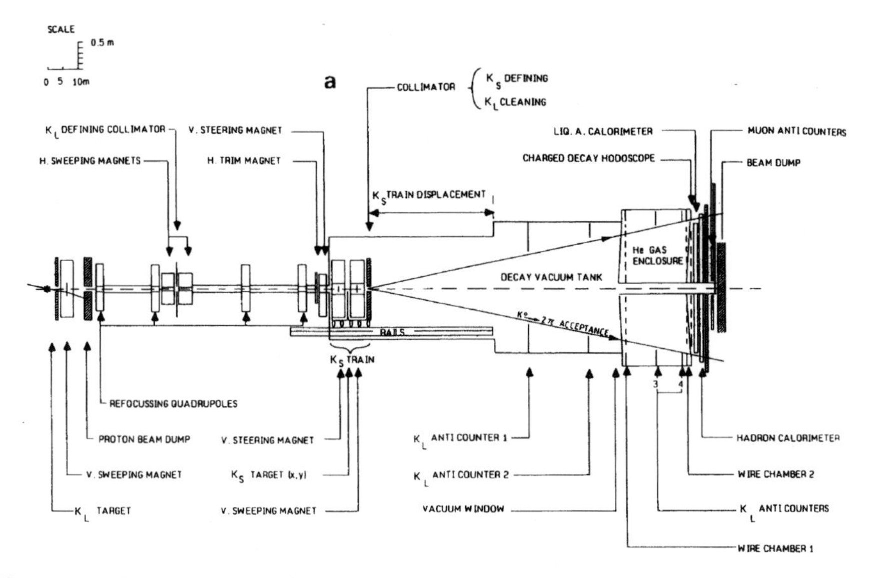

Fig. 7.12. Layout of the CERN experiment NA31 to measure the double ratio $R = |\eta_{+-}/\eta_{00}|^2 = [(L \to +-)/(S \to +-)]/[(L \to 00)/(S \to 00)]$, and that gave first evidence for direct CP violation. N.B. the horizontal and vertical scales differ by a factor of 20

The hadron calorimeter consisted of a multi-sandwich of 2.5-cm-thick iron plates interleaved with planes made of plastic scintillation counter strips.

The more splendid parts of the detector were the electromagnetic calorimeter and the short-lived beam. The former was a lead-liquid argon sandwich calorimeter. The lead serves to generate the shower, and the ionization produced by passage through the inert liquid is collected and measured to give the energy contained in the shower. Each of the 80 sandwich layers consisted of a sheet of lead, 1.5 mm thick, followed by a liquid argon–filled gap, 2 mm thick, a plastic readout plane, 1 mm thick, plated on each side with copper in a strip pattern, the strips, alternately horizontal and vertical, 12.5 mm wide. The assembly was in a vacuum-insulated vessel cooled below the $-186\,°\mathrm{C}$ boiling point of the argon. The detector excelled in space resolution, about 1 mm for the center of a photon shower, as well as in energy resolution, about 1.5%, for the dominant photon energies of about 20 GeV. Another special feature was the K_S beam train. Since the K_S decay length was much shorter than the length of the decay region, and since the detection efficiency is a function of position, it was necessary to measure the K_S decay, with the beam-producing target at several positions along the beam axis. To this end a train was built that could roll the target and the associated steering and focusing magnets of the proton beam along the beam axis, so that the $\pi^+\pi^+/\pi^0\pi^0$ ratio could be measured also for the short-lived kaon for the whole K_L decay region.

The statistical error for this experiment is dominated by the observed number of long-lived neutral decays. This was 10^5, three orders of magnitude larger than the numbers observed in the experiments of the 1970s. It was not trivial to confine the systematic uncertainties to the new statistical precision, and in the end they were estimated to be somewhat larger. The first published result [103], $\mathrm{Re}(\epsilon'/\epsilon) = (3.3 \pm 1.1) \times 10^{-3}$, showed, with three standard deviation certainty, that there is direct CP violation.

The experiment continued, with some improvements, in particular, a transition radiation detector following the spectrometer. Unfortunately, I had to discontinue my own participation, since I was heavily engaged in the preparation of an experiment for the LEP electron–positron collider, under construction and nearing completion. The second publication [104] of the group managed to reduce significantly both systematic and statistical errors, with the result $\mathrm{Re}(\epsilon'/\epsilon) = (2.0 \pm 0.7) \times 10^{-3}$. The combined CERN result was $\mathrm{Re}(\epsilon'/\epsilon) = (2.3 \pm 0.65) \times 10^{-3}$.

The Fermilab experiment [105] differed in several ways. All four decay rates were measured simultaneously, a clear systematic advantage; however, the decay region for the long-lived decays was only partially covered by the neutral beam, so the relative acceptances for charged and neutral decay for the remainder of the decay region were not measured, but had to be calculated. The neutral beam was regenerated and therefore contained a small diffractive component that required correction. The detector contained a magnetic spectrometer so that the momenta of the charged particles were measured, and, finally, the electromagnetic calorimeter used the Cherenkov radiation produced by the showers in a wall of 804 lead glass blocks, 6 cm on a side, with consequently poorer space resolution than the CERN experiment. The result, $\mathrm{Re}(\epsilon'/\epsilon) = (0.74 \pm 0.60) \times 10^{-3}$, was consistent with no direct CP violation.

Both groups went on to prepare a new round of even more beautiful experiments to measure the double ratio R. Both were able to improve the speed of the electronics so that the data-taking rate could be increased by a factor of almost ten. The Fermilab experiment, now with almost 100 collaborators from 13 institutions, was basically along the lines of the previous experiment, with the most notable improvement the electromagnetic calorimeter, essentially replacing each lead glass block with four cesium iodide fluorescing crystals, thereby improving space and energy resolutions. The new CERN experiment, NA48, with 150 collaborators (the author unfortunately not among these) from 16 institutions, made three major changes. There were now simultaneously two beams so that the four rates could be measured simultaneously. In addition to the K_L beam, a K_S beam was produced by a small proton flux on a target a few meters upstream of the decay region, laterally about 10 cm from the K_L beam. It was directed to intersect with the K_L beam at the electromagnetic calorimeter. K_S events could be tagged by a signal produced by a counter in the proton beam. The decay region used in the analysis included only that portion, 3.5 K_S lifetimes in length, adequately covered by the K_S beam. The charged particle tracking now included a magnetic spectrometer. Technically the most demanding improvement was the new electromagnetic calorimeter. Liquid krypton replaced both the lead and the liquid argon of NA31. The electrode structure consisted of copper-beryllium ribbons stretched along the beam direction on a 2×2 cm lattice in the transverse plane, altogether 13,000 readout cells. The krypton was maintained in a cryostat below its boiling point of $-157\,^{\circ}$C, and much of the electronics had to be incorporated in this cryostat in order to achieve the required time resolution. Both experiments are now completed. The final results for ϵ'/ϵ $\mathrm{Re}(\epsilon'/\epsilon) = (1.47 \pm 0.22) \times 10^{-3}$ for CERN [106] and $\mathrm{Re}(\epsilon'/\epsilon) = (2.07 \pm 0.28) \times 10^{-3}$ for Fermilab [107].

It would now be most interesting to compare this more precise result with the expectation of the Standard Model, contained in the K-M mixing matrix, which was mentioned above. Although in principle this prediction is well-defined, the calculation encounters technical difficulties, which as yet are not entirely resolved, so that there is a correspondingly large uncertainty in this prediction. Present calculations fall, more or less, into the range $(0.5 < \mathrm{Re}(\epsilon'/\epsilon) < 2) \times 10^{-3}$, of the same order as the experimental result. Despite the lack of precision, this is an important indication that the Standard Model does encompass the CP violation observed experimentally. There is now a very large experimental effort to observe certain CP violation effects in the decay of B^0 mesons. The B and $\overline{B}^0$ mesons are composed of bottom and antidown quarks, and antibottom and down quarks, respectively, and are about ten times heavier than the K^0 mesons. Colliders have been specially constructed for these experiments, one in Japan and one in the U.S. One of the great motivations for these experiments is that, besides providing checks on the implications of the CP violation experiments so far entirely confined to the K^0 system, some of the expected effects in the B^0 system are more amenable to calculation in the Standard Model so that the adequacy of the model can be be tested with greater precision.

Both experiments have observed clear CP violation with the rare decay channel (decay fraction $\sim 1/1\,000$) : $B^0 \rightarrow \Psi + K^0$. The $B^0 - \overline{B}^0$ system is similar to the $K^0 - \overline{K}^0$ system, the particles of well-defined lifetime are mixtures of the B^0 and $\overline{B}^0$ states of well-defined particles or antiparticle (charge conjugation) property. Unlike the neutral kaon decay, the CP violation expected for the above decay channel is not small, but of order unity. The observed CP violation can be quantified by the parameter $\sin 2\beta$. The results for the two experiments are:

Belle (Japan) [108] $\sin 2\beta = .72 \pm .08$
BaBar (U.S.) [109] $\sin 2\beta = .74 \pm .08$

These results are in agreement with the Standard Model expectation of $\sin 2\beta = .72 \pm .07$, where the theoretical uncertainty reflects our inability to calculate certain consequences of the strong interaction, although in principle these are well-defined in the Standard Model.

Even if we assume that the Standard Model does *describe* the observed CP violation, it is not really a theory of these effects. CP violation is just introduced, ad hoc, as a parameter. A more profound understanding is essential. Perhaps the fundamental understanding of CP violation is intimately connected with that of "flavour symmetry." At present we know that we have three "families," and the corresponding six quark flavors, and three lepton and neutrino flavors, but the underlying relationship between the three families is not known. It is possible that the understanding of flavor symmetry will also permit an understanding of flavor mixing, as well as CP violation in the weak interaction.

8

Neutrinos, II

8.1 Quarks and Gluons

The year 1969 witnessed a breakthrough in our understanding of nucleons and nuclei. The experiment underlying this breakthrough was performed on the newly constructed two-mile-long linear accelerator at Stanford University, which produced beams of electrons of up to 20 GeV. For the first time it was possible to study highly inelastic collisions of the electrons on nucleons, in which the scattered nucleon was accompanied by newly created, generally unstable, hadronic (strongly interacting) particles such as pions and kaons. In the experiment [110], only the outgoing electron was detected, and the scattering probability, as a function of its energy loss and the angle by which it was scattered, were measured.

Figure 8.1 shows the most striking result: The event rate was found essentially independent of Q^2, the square of the momentum lost by the electron. This behavior, called "scaling," is the expectation for collisions with point-like objects, and it could be inferred that the nucleon is not elementary but composed of constituents, which scattered as if they were point-like objects, that is, elementary particles. These were at first called "partons" by Feynman, one of the first to appreciate the dramatic implications of the result, although Bjorken had anticipated it two years earlier. These partons are now the "quarks" and "gluons" of our "Standard Model". The proton is made up of two up and one down quark, for the neutron it is two down and one up quark, bound together by the strong interaction, propagated by the massless and electrically uncharged gluons. Now we know that there are three families of particles, each containing a charged lepton, a neutrino, and two quarks; an up-type quark of electric charge $+2/3$ (the charge of the positron); and a down-type quark of charge $-1/3$ (see Table 6.1 at the end of Chap. 6).

Three of the quarks: the up and down quarks, as well as the strange quark, invented to understand the strange particles such as the kaon and Λ hyperon, had been anticipated in the early sixties by Gell-Mann and Zweig. The great Stanford discovery was followed in the late sixties, early seventies by our

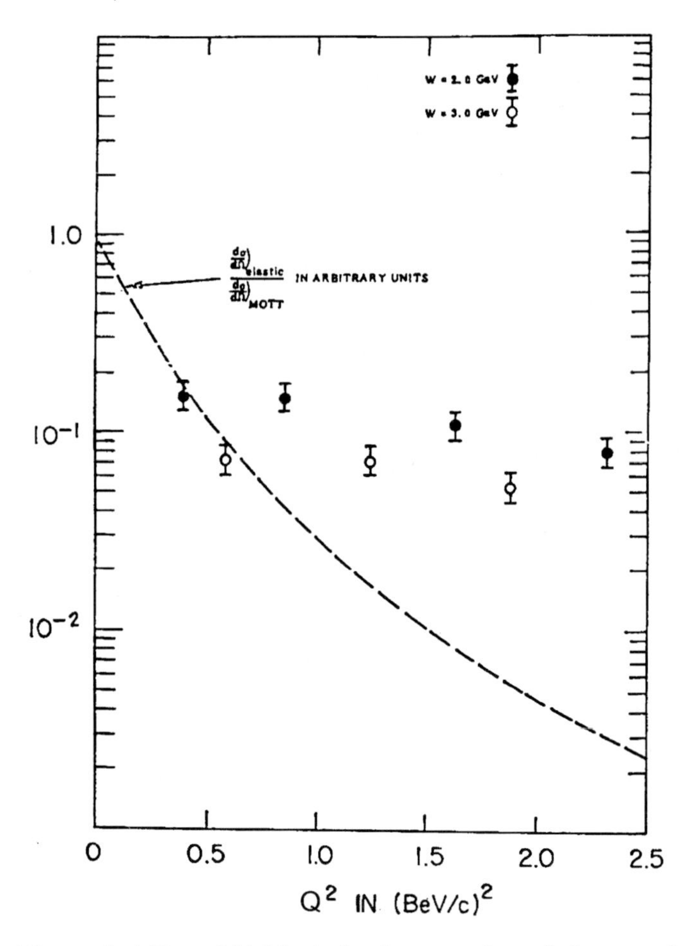

Fig. 8.1. The probability of highly inelastic scattering of electrons by protons, plotted as function of the square of the momentum transferred to the nucleon, Q^2. It is seen to be essentially independent of Q^2, as expected if the nucleon were composed of point-like objects [110]

first theory of the strong interaction, "Quantum Chromo Dynamics (QCD)," in which each quark exists in three varieties, called "color," and the gluon in eight color combinations. QCD has a gauge symmetry as does the Electroweak theory, so there is a widely held belief that there is an underlying "Grand Unified" theory that incorporates both, but has yet to be discovered. The other particles of the Standard Model were discovered in the following three decades:

1973 The charm quark, at Brookhaven National Laboratory and Stanford, in two independent experiments.
1974 The tau meson, at the Stanford electron–positron collider.
1977 The bottom quark, at Fermilab.
1979 The gluon, partly as a confirmation at CERN of quantitative predictions of QCD, partly by the observation of effects of gluon radiation in electron–positron collisions at DESY, Hamburg.
1983 The weak, heavy bosons W^+, W^-, and Z^0, at CERN, at the 550 GeV proton-antiproton collider.
1994 The top quark, at Fermilab, at the 2,000 GeV proton-antiprotoncollider.
2000 The tau neutrino, at Fermilab, in a nuclear emulsion detector exposed to neutrino decay products of very short-lived particles, produced by 1,000 GeV protons in a dense target.

So, finally, all the particles of the three families, as well as the gauge bosons responsible for their interactions, have been seen. Moreover, all that is known now about elementary particles and their interactions, and much of this with great precision, is understood in the frame of the Standard Model, that is, the Electro-Weak and QCD theories. But, and very important, this does not mean that we know everything. There are many things missing in the Standard Model, such as any understanding of the particle masses and of the interaction strengths. A striking element of the theory that remains to be confirmed is the existence of a particle postulated in the theory and, essential to its coherence, a scalar (spin zero) particle called the Higgs. We will come back to this problem. But it is already remarkable to have so much that fits together in a coherent theory, beautiful progress since the time when I was a young physicist just after the war.

8.2 CDHS

But back to our group in CERN in 1972. We were about ten, happily measuring CP violation in the K^S–K^L system, and other properties of neutral kaons as well as the Λ^0 and $\sum^0$ hyperons. In the meantime, at Batavia, IL, near Chicago, a new laboratory, Fermilab, was under construction with, as central resource, a machine to accelerate protons to 400 GeV, more than ten times as energetic as then-existing accelerators. This started to operate in 1973. Europe, after long disputes about where it should be sited, in 1972 agreed to build a similar machine at CERN, which would become operational in 1976. Given the Stanford discovery of the composite nature of the nucleon, it was evident that neutrinos, not encumbered with the complex strong interaction, would provide an excellent tool to study the nucleon structure. Our group was joined by Rene Turlay, co-discoverer in 1964 of CP violation, and several of his collaborators at the Saclay Laboratory near Paris to plan an experiment and an appropriate detector to study the interactions of the higher energy neutrinos, which would become available, with nucleons. CERN decided to construct one neutrino beam in which four experiments would be lined up, one behind the other, laid out so that it would traverse the Big European Bubble Chamber, already in place and not easy to move. Behind this would follow our experiment, called CDHS for its collaborating institutions, CERN, Dortmund, Heidelberg, and Saclay followed by another electronic detector named CHARM for CERN, Hamburg, Amsterdam, Rome, and Moscow; and finally Gargamelle, the freon-filled bubble chamber of neutral current fame. As it turned out, Gargamelle developed a mechanical fault and was never operated in this beam. The neutrino beam consisted of a metal target in which the protons produced pions and kaons, a 150-m-long section in which the secondary beam could be focused, a 300-m-long evacuated decay tube, and a 400-m-long earth shield to filter out particles other than neutrinos.

Our CDHS detector was made of circular iron plates, 3.75 m in diameter, in 19 modules, each weighing 65 tons. The first seven modules consisted each of 15 plates, 5 cm thick, the last 12 modules of five plates, 15 cm thick. Each module was equipped with copper windings to produce a torroidial magnetization of 15 kilogauss. In the gaps between the iron plates, plastic scintillators were inserted to measure the particle fluxes, and between modules, drift chambers were placed to track the trajectories of traversing particles. The iron plates served, firstly, as a massive target in which the neutrinos could produce their interactions. Secondly, they served to degrade the hadrons produced in these interactions; the resulting showers of particles were contained within about a meter of iron, and their energies could be determined from the total light produced in the scintillators. Thirdly, the magnetic field served to bend the trajectories of the only penetrating particles, the muons, so that their energies, proportional to the radius of curvature of the trajectory, could be obtained from the wire chamber measurements. Figure 8.2a shows the stringing, at Saclay, of one of the drift chamber planes. Each wire chamber consisted of three such planes, with wires at 60° to one another. Figure 8.2b is a photo of the two electronic neutrino detectors, CDHS and CHARM, installed in the neutrino beam.

There is a historical aside here that I have always considered interesting. Our original plan for the detector was substantially different. It consisted of two parts. The front part served as neutrino target and hadron shower energy measuring instrument, and this was followed by an iron plate magnet with interspersed wire chambers to measure the muon momenta. However, the management did not like our front part and proposed that we do this experiment, combining our magnet with a target-hadron calorimeter part provided by another team. This was not exactly to our liking; we were not convinced of the other proposed method, and we preferred to be independent. Only then did we notice that the magnet could be transformed to do all three functions simultaneously, with an overall improvement in the detectors capability. I remember this as an illustration of a) how limited our vision can be – why did

Fig. 8.2. **a**) The stringing of a drift chamber plane at the Saclay laboratory near Paris. **b**) The CDHS and Charm detectors in the neutrino beam line at the SPS at CERN, 1977 (with permission of Jack Steinberger)

we not see this in the first place? – and b) how bad luck can turn to one's advantage.

The neutrinos of the beam, as before, are muon neutrinos. The basic reactions for the inelastic scatterings with which we are concerned are:

a) Charged Current (CC): neutrino + nucleon → muon + hadrons (from the decay of the exited hadron),
b) Neutral Current (NC): neutrino + nucleon → neutrino + hadrons.

In the quark model of nucleon structure, this reads:

a) Charged Current: neutrino + quark → muon + exited quark, and
b) Neutral Current: neutrino + quark → neutrino + exited quark.

An excited quark is observed as a jet of hadrons.

Usually the details of the hadron jet is of little interest, and an event is characterized by three measurable parameters: 1) the "effective" mass fraction of the nucleon, carried by the struck quark, called x, 2) the fraction of the neutrino energy transferred to the quark, called y, and 3) the square of the neutrino momentum transferred to the quark, Q^2, which we encountered already in connection with the Stanford experiment. The dependence of the scattering probability on y and Q^2 was predicted by the quark-parton model. In particular, the scattering probability was expected to be independent of Q^2, and, as we already saw, this was the great Stanford discovery of 1969. The dependencies on x, called the "structure functions," describe nucleon structure, how the different constituents are distributed in the nucleon. These structure functions cannot be predicted by the quark-parton model,[1] but are the same functions for electron and neutrino inelastic scattering. Particular structure functions describe the x distributions of the different constituents of the nucleon. There are the "valence quarks," that is, the three basic quarks of the nucleon, two up and one down for the proton, one up and two down for the neutron, then there is the "sea" of quark-antiquark pairs, then there are gluons, which, however, don't interact with either electrons or neutrinos so that the gluon distribution cannot be directly measured.

8.3 Nucleon Structure

In Stanford, following the scaling discovery, the total structure function describing the total quark + anti-quark, valence + sea quark distribution was measured. In 1974, in yet another beautiful, important result, the Gargamelle bubble chamber group at CERN could show [111] within the rather substantial errors, that the structure function measured in neutrino and anti-neutrino scattering has the same shape as that measured in electron scattering, and differs in magnitude by the factor 18/5, which the quark-parton model predicted (Fig. 8.3).

This factor 18/5 just reflects the quark electric charges. For a nucleus with equal numbers of protons and neutrons it is: neutrino/electron $= 2/[(2/3)^2 + (-1/3)^2] = 18/5$, where 2/3 and $-1/3$ are just the electrical charges of up and down quarks.

In 1973, while our own 1,600-ton electronic detector was under construction, the Fermilab accelerator came into operation, not without some difficulties, and the three neutrino experiments there, two electronic and a bubble chamber similar to the Big European Bubble Chamber at CERN, began to

[1] In principle, the structure functions may be calculated in Quantum Chromo Dynamics, QCD, the theory of the strong interaction, part of the Standard Model, but actual calculations can only be carried out numerically and seem to exceed present calculational possibilities.

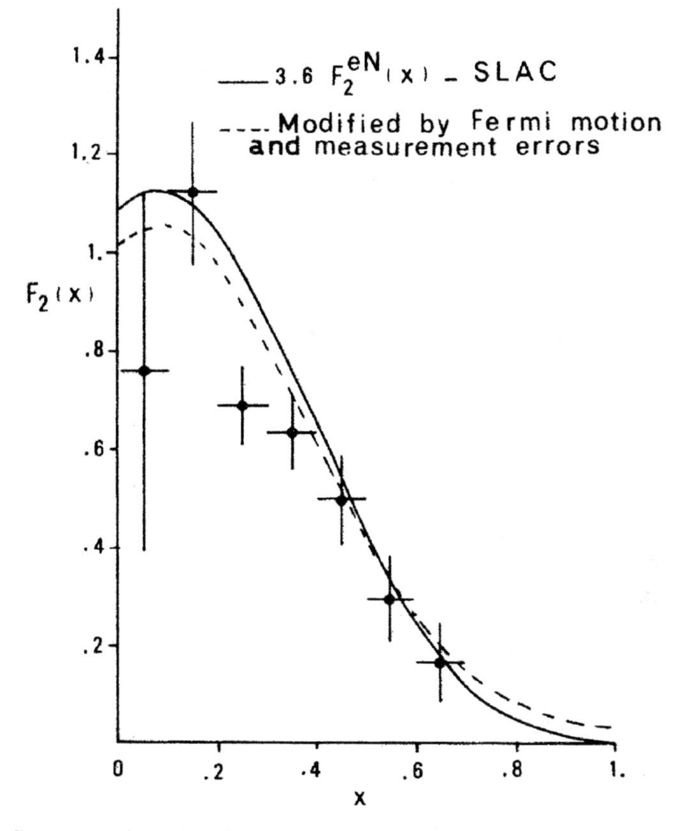

Fig. 8.3. Structure function for the combined valence + sea quark distribution, Gargamelle neutrino results (*crosses*) [111], compared with the electron results (*solid lines*), previously measured at Stanford, and multiplied by the factor 18/5 expected in the quark-parton model

obtain results. They confirmed, not without some oscillations but eventually with greater clarity, the Gargamelle neutral current discovery and extended the charged current result to higher energy, and with much improved statistical accuracy. Perhaps the most dramatic results were those of the Harvard-Princeton-Wisconsin-Fermilab group, which claimed discovery of several deviations from the predictions of the quark-parton model. The authors named the deviations the "high y anomaly."

At CERN, the SPS became operational in the fall of 1976. The neutrino event rate was many thousands per day, compared to the one per day in the 1962 Brookhaven experiment. The experiment operated about seven years, we were a team of about 30, and with the help of these neutrinos, a good deal could be learned, both about the Standard Model and the structure of the nucleon. Figure 8.4 shows a charged current event and Fig. 8.5 is the

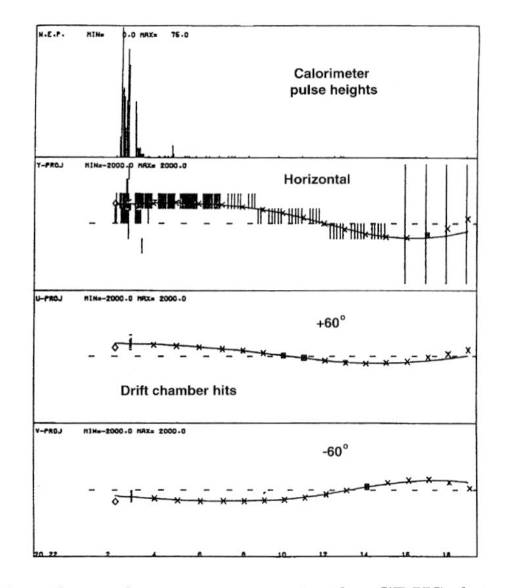

Fig. 8.4. A neutrino-charged current event in the CDHS detector. The neutrinos enter horizontally from the left. A neutrino interacts in the second of the 19 modules. The top panel shows the energy deposited by the hadron shower in the scintillators between the iron plates, the other three views show the wire chamber positions of the muon track in the three projections. The line is the computer-reconstructed muon track. For neutral current events, the muon track is missing, but the hadron shower is seen, and so the hadron energy measured. They were distinguished from the charged current events by the missing track

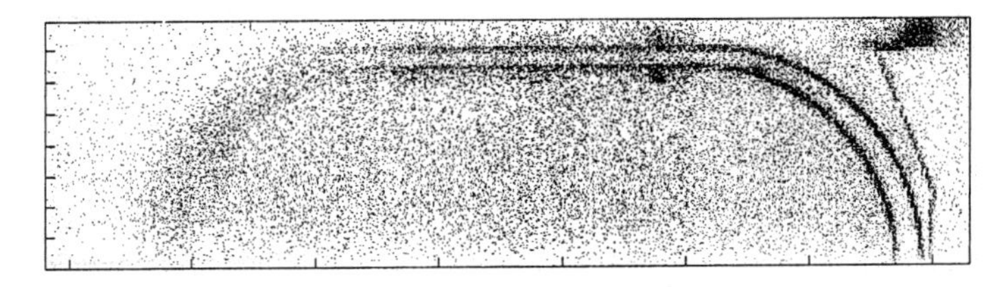

Fig. 8.5. Neutrino "X-ray" of a hydrogen dewar. In the later stages of the experiment we measured the interaction of neutrinos with protons (rather than iron), putting a hydrogen-filled tank (dewar), followed by wire chambers, in front of the detector. The reconstruction of the origin of the neutrino events, on the basis of the tracks measured in the wire chambers, produced this one and only neutrino "X-ray"

neutrino "X-ray" of a liquid hydrogen–containing dewar tank put in front of the detector to permit the measurement of neutrino interaction with protons.

The first important result, obtained within a few months, showed that the Fermilab "high y anomaly" results were erroneous; in fact, the parton model described our more precise results correctly [112]. For the weak neutral current, it was possible to measure the ratio R of neutral-to-charged rates, a measure of the so-called "weak mixing angle," θ_w, an important parameter in the electro-weak theory, with higher precision. And for the first time, it was possible to measure the event rate as a function of the energy transmitted to the hadrons and to compare this with the predictions of the electro-weak theory [113].

In neutrino interactions, the valence and sea quark distributions can be determined separately by taking sums and differences of the neutrino and antineutrino event rates, multiplied by appropriate factors of the energy transferred to the nucleon. Some CDHS results [114] are shown in Fig. 8.6.

Among the sea quark-antiquark pairs, heavier quarks are also expected to be present, but, because of their higher masses, in much smaller amounts. The most abundant is expected to be the strange quark, with some 20 times the mass of the up and down quarks. Very small charm or top quark contributions are expected, because of their much higher masses. It was possible to observe the strange quark-antiquark sea by measuring the rates for events in which two oppositely charged muons were present. Such events, occurring several hundred times less frequently than the usual, single muon events, could be attributed to the collision of the neutrino with a strange quark, changing this to a charmed quark. The emitted charmed hadron decays within a very short time, producing a muon with a probability that is known. In this way, it can be seen [115] that the strange sea structure function is of very similar shape as the up-down sea, but with relative magnitude, s/(u+d), 20 times smaller.

8.4 Quantitative Confirmation of "Scaling Violations" Predicted by QCD Theory

Probably the most important result of the SPS neutrino experiments was the observation of "scaling violations," providing a first quantitative confirmation of the new theory of the strong interaction, QCD. The scaling property, that is, the independence of the structure functions with respect to momentum transfer squared, Q^2, as already noted, was the key observation leading to the recognition, at Stanford in 1969 [110], that nucleons are compound objects. However, the more detailed, quantitative theory of the strong interaction, the gauge theory of the interaction of quarks with gluon vector bosons, which followed in 1973, predicted certain deviations from scaling. Although the QCD theory was theoretically very attractive at the time, both because it provided a theoretical basis for the quark-parton model and because its structure was formally very similar to that of the successful electro-weak theory, there was no quantitative prediction of the theory that had as yet been confirmed. The CDHS experimental results [116] are shown in Fig. 8.7, where the Q^2 dependence is plotted for the valence structure function, in bins of x. If scaling were not violated, the experimental points would lie on horizontal lines. However, the experiment showed that for small x the values increase with Q^2, and for large x they decrease. The theoretical predictions for the magnitude of these

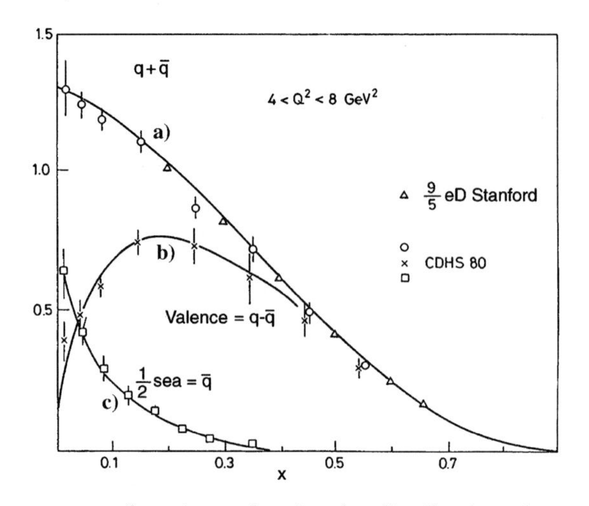

Fig. 8.6. The structure functions, that is, the distributions in x, **a**) for the sum of quarks + antiquarks in the nucleon, obtained from the sum of neutrino and antineutrino scattering, **b**) the distribution in x of the "valence" quarks, that is, the difference, quarks – antiquarks, obtained from the difference between neutrino and antineutrino scattering, and **c**) the distribution in x of the "sea" quarks, obtained from the difference of these two distributions. The parameter x can be thought of as the fraction of the nucleon energy carried by the struck quark

scaling violations are proportional to the strength of the strong interaction, a free parameter in the theory, measured in this experiment for the first time. Given the strength, the shapes were predicted by the theory. The agreement of the predicted shapes with the observation thus gave experimental support to the new theory, a milestone in the establishment of the Standard Model.

An important fact, known already from the Stanford determination of the structure function for the combination of quarks and antiquarks in the early 1970s, was that the total nucleon mass carried by the sum of quarks and antiquarks, as measured by the structure functions, accounts for only one half of the nucleon mass. The remaining half is presumably in the form of gluons, which do not interact with either electrons or neutrinos. The gluon structure function could not be measured directly, but in the QCD theory the gluons are directly related to, in equilibrium, with the quark-antiquark sea: a gluon can create a quark-antiquark pair, and a quark and antiquark can annihilate each other, producing gluons. The theory links the gluon structure function to the quark structure functions, as well as to their scaling violations. Once these were known, it was possible to invert them to get the gluon distribution [117]. The first result for the distribution of the gluons in the nucleon is shown in Fig. 8.8.

This inversion of the quark structure functions to get the gluons was a nontrivial accomplishment, led for us by Franz Eisele.

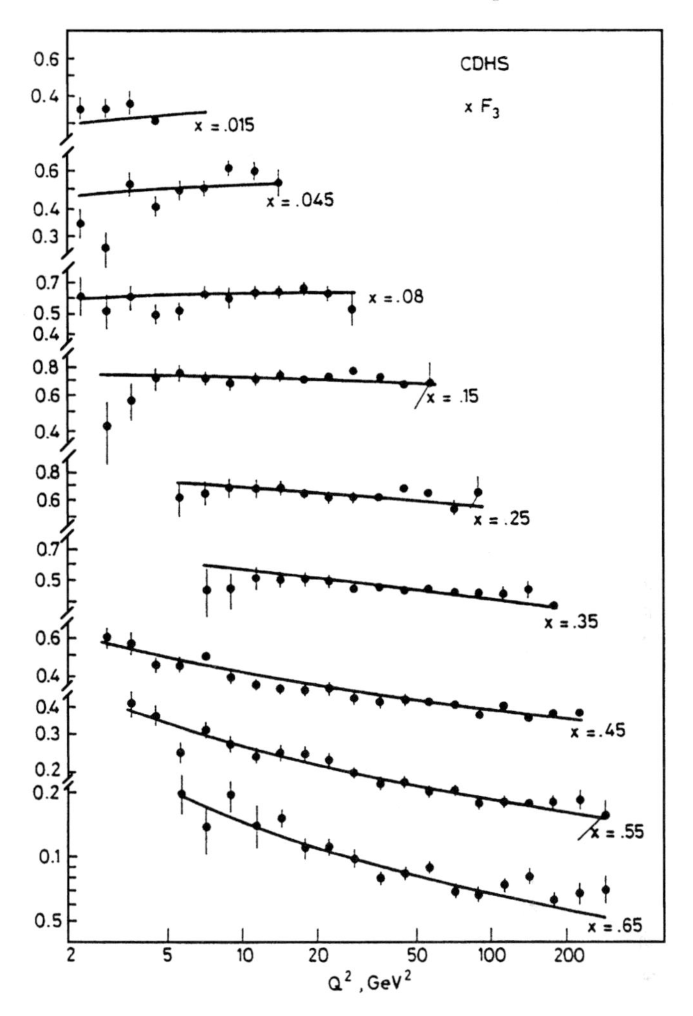

Fig. 8.7. Demonstration of "scaling violation," that is, the variation of the structure function, in this case, the structure function for the valence quarks, with momentum transfer squared, Q^2, for different values of x. In the simple quark-parton model, no variation with Q^2 is expected. The experiment confirms the predictions of the more sophisticated QCD theory, the Q^2 variations shown as *solid lines*

8.5 Beam Dump and Tau Neutrino

In this experiment, an absorber was placed directly following the target in which the SPS protons produced the pions and kaons, whose decay, in turn, produced the neutrinos of the neutrino beam, so this was an experiment without a neutrino beam. The hope was to detect the tau neutrino, which was needed in the Standard Model to go along with the tau lepton, but which had not been seen. The 400 GeV protons were sufficiently energetic to produce,

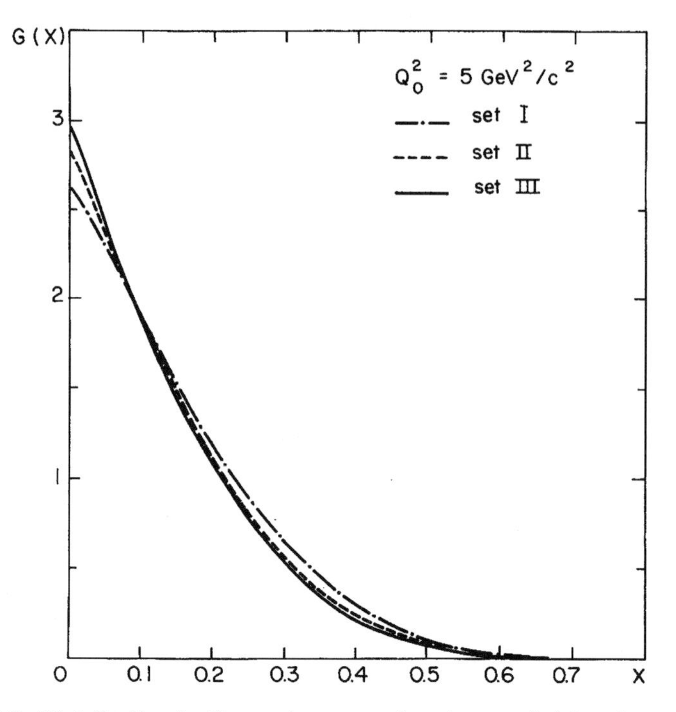

Fig. 8.8. Distribution in the nuclear mass fraction carried by gluons [117]

with a probability of some fraction of one percent, mesons containing a bottom quark. These, in turn, decay very quickly, within a millimeter, while still in the target and produce tau neutrinos a few percent of the time. So a tau neutrino flux was expected, but of low intensity, lower than the flux of muon neutrinos due to pion and kaon decay, before these could be stopped in the absorber. The events expected in the CDHS detector from tau neutrinos were somewhat different than those produced by muon neutrinos, but in the end we could not be sufficiently sure that some particular events observed were due to tau, rather than muon neutrinos. The tau neutrinos were finally discovered, the last of the Standard Model particles at Fermilab in the year 2000 in such a beam dump experiment, but with a detector using nuclear emulsions, which have sufficiently high space resolution so that the very short, but finite, lifetime of the tau lepton, produced in the tau neutrino reaction, could actually be observed.

8.6 Neutrino Masses, Neutrino Mixing, Neutrino Oscillations

One of the most interesting aspects of neutrino behavior is that of neutrino mixing and oscillations. This is a quantum mechanical effect intimately connected with possible masses of the neutrinos. The masses of neutrinos were known to be very much smaller than those of other fermions, and it was convenient to imagine that neutrinos are massless. But if they do have nonzero masses, a new field of physics opens up, that of the oscillation of neutrinos of one flavor to another. This was first anticipated by Pontecorvo [118], the enormously inventive Italian, Canadian, and Soviet physicist whom we already encountered in connection with the search for the muon neutrino. It was he who first suggested that one might look for neutrino mixing and oscillation. The basic idea is that a neutrino "state" with well-defined "flavor," say, a muon neutrino, originating in muon decay, is not a state of well-defined "mass." But it is the mass of a state that characterizes the propagation with time. Consider a system of two types of neutrinos. If the states of well-defined mass are not identical with the states of well-defined flavor, then each of the two states of well-defined flavor can be thought of as consisting of a mixture of the two mass states, with the relative amplitudes of the two states, the so-called "mixing angle," a parameter. The system is then described by three parameters: the two masses and the mixing angle. For a system of three neutrinos there are three masses and four mixing angles. If the mixing angle is zero, the state with well-defined flavor is identical with a state of well-defined mass, there is no mixing, and the neutrino will retain its flavor with time, but, otherwise, there is mixing and oscillation. With time the neutrino flavor character changes, a neutrino born with well-defined flavor becomes a mixture of the two (or three) flavors. The frequency of the oscillation is 1/2 * (square of the difference in mass)/(neutrino energy), in units in which Planck's constant and the velocity of light are unity. This corresponds to a wavelength for neutrino mixing of 2.5 km * (neutrino energy in GeV)/(difference in the squares of the two masses in eV^2). The maximum fraction of the new flavor, in the case of two flavors, is the square of the sine of twice the mixing angle.

The first searches for neutrino mixing were conducted at nuclear reactors, which are copious sources of electron antineutrinos. Basically, detectors capable of detecting and measuring the flux of electron antineutrinos are placed in two positions: one close by, the other further away. If the neutrinos do not oscillate, the observed fluxes will be in the inverse of the squares of their distances from the reactor. If, however, the electron neutrino has in part oscillated to some other flavor, a smaller amount will be detected in the farther detector. Early searches were negative. No oscillations were observed, permitting the conclusion that either the mixing angle is smaller than one degree, or the difference in the squares of the masses is smaller than 0.01 eV^2.

At CERN, in 1983, we looked for possible oscillations of our muon neutrinos. To increase the sensitivity for lower neutrino masses, a lower energy beam was built, using the PS proton beam, which produced a neutrino beam with average energy of about 2 GeV. By this time, CDHS had been augmented by 12 improved modules with higher spatial resolution of the hadron shower. A forward detector was assembled with six modules; the bulk of the detector, the 21 remaining modules, was left in place, some 900 m behind the front de-

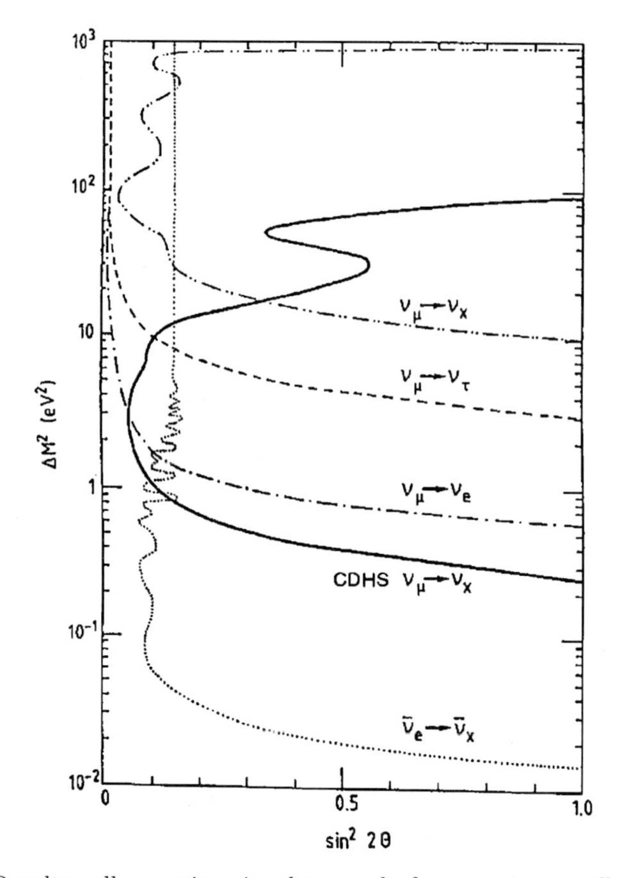

Fig. 8.9. Results, all negative, in the search for neutrino oscillations in the $\Delta m^2 \sin 2\theta$ plane. The vertical axis is the (logarithm of) the difference in the square of the masses, the horizontal axis is the square of the sine of the mixing angle. The CDHS experiment excludes the area to the right of the *solid line*, for the oscillation of the muon neutrino to any other flavor. The dotted curve is the result of the most sensitive reactor experiment at that time. The parameters to the right are excluded for the oscillation of electron neutrinos to any other flavor. The *dot-dash curve* is the result of the search in the Big European Bubble Chamber for the electron signal from an initial mu-neutrino, the *dashed curve* shows the area excluded in a search at Fermilab for a tau-neutrino signal from an original muon neutrino beam

tector. We looked for a muon neutrino signal, that is, events with secondary muons. Oscillation would have produced a diminution of this signal in the back detector, but no such decrease could be detected [119]. These negative results are shown in Fig. 8.9, together with several other accelerator results of that time, all negative.

But neutrino masses, mixing, and oscillations have now been seen, with the help of neutrinos not from accelerators but from the sun and cosmic rays.

Let us start with the sun. The nuclear energy production process in the sun requires, as a first step, the combination of two protons to form a deuteron. In this reaction, a neutrino and a positron are emitted. In the following reactions, in which deuterons combine to make helium and higher elements, additional neutrinos are produced in the decays of the different nuclei that are formed. The whole process has been the subject of extensive studies and is quite well understood. This "solar model" predicts the neutrino fluxes from the different decays whose energy spectra are known. Since the total energy production in the sun is known, there is a known normalization of the neutrino flux. About two percent of the energy radiated by the sun is in the form of neutrinos, corresponding to a flux of $\sim 10^{10}$ neutrinos per cm^2 per second on the surface of the earth. The bulk of the neutrino flux predicted by the solar model is from the primary p-p process, with neutrino energies 0.4 MeV or less. Then there is a mono-energetic flux, with about ten times smaller intensity, at 0.9 MeV due to Be^7 decay and a roughly 1,000 times smaller flux due to B^8 decay but with much higher energies, up to 15 MeV. The first attempts to detect solar neutrinos date back to the seventies.

The experiments are very difficult, because of the extreme reluctance of the neutrinos to interact. Three detecting systems have been successfully used, each sensitive to different portions of the solar neutrino energy spectrum. The first method uses the excitation, by the neutrino, of an isotope of chlorine, Cl^{37}, to an isotope of argon, A^{37}, a reaction initially suggested by Pontecorvo in 1946! The experiment is still running; there are 130 tons of chlorine in tanks deep underground. Solar neutrinos produce about .4 atoms of Argon per day, and these are fished out by bubbling a very small quantity of normal argon through the tanks once a month. Then they are detected by their radioactive decay. The reaction is sensitive to the part of the neutrino spectrum above .81 MeV. A similar experiment, in operation since about a decade ago, uses, instead of the conversion of Cl^{37} to A^{37}, the conversion of Ga^{71} to Ge^{71}. There are two versions: a European collaboration, installed in the Grand Sasso tunnel in Italy, and a Soviet-U.S. collaboration, somewhere in the former Soviet Union. The energy threshold for this reaction is 0.21 MeV: It is the only experiment sensitive to the main neutrino flux from the sun, that from the p-p reaction. Finally, there is the experiment in the Kamioka mine in Japan. The detector is just a very big tank of clear water. The present (second) version is 40 m in diameter and 40 m high, surrounded by 10,000 very large photo-tubes, which detect the light produced by charged particles in the water (Fig. 8.10). This light is directional, and the pattern of light produced in the photo-tubes permits reconstruction of the direction of the track as well as its energy and location in the detector.

The detector shown in Fig. 8.10 is Super-Kamiokande, the second, even larger and more beautiful version of an already very beautiful detector, Kamiokande. Kamiokande was originally constructed to look for the possible but surely very rare decay of the proton, but no evidence for proton decay was found. Instead, Kamiokande and Super-Kamiokande played a leading role

Fig. 8.10. Inside view of the Super-Kamiokande water tank, viewed by 10,000 50-cm-diameter photo-tubes which cover its walls, before being filled with water

in the discovery of neutrino oscillations, a no less fundamental phenomenon in particle physics – a kind of physics which those who conceived Kamiokande did not have in mind at all. Super-Kamiokande has, at the moment of this writing, operated for several years with magnificent results, helping us to understand the phenomenon of neutrino oscillations. However, since a few months ago, it is in shambles. During an operation of clean-up and replacement of the water in the spring of 2002, one of the photo-tubes imploded during the refill. The resultant shock wave destroyed the remaining 6,000 tubes then under the water. Super-Kamiokande will be rebuilt, after it is clear how a future accident can be avoided, and we look forward to more excellent results from it. I tell

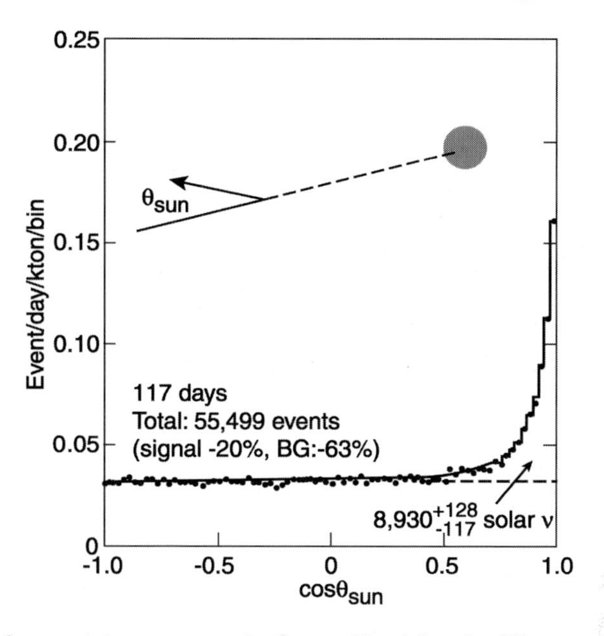

Fig. 8.11. Solar neutrinos as seen in Super Kamiokande. The number of events plotted as function of the cosine of the angle of the electron track with respect to the sun. The peak on the right (small angles) is produced by neutrinos from the sun. The attenuation of the neutrinos in passing through the earth is negligible so that the sun is seen equally well day and night

these two stories because they are such good illustrations of the role chance plays in our lives as researchers.

I have always regarded this detector as particularly simple and beautiful, and I was pleased with the important results obtained from it and in the fact that the experiment and its originator, Masatoshi Koshiba, shared the 2002 Nobel Prize.

The detection of the solar neutrinos here is through their interaction with the atomic electrons. The electron is scattered forward. Its direction does not differ very much from that of the incident neutrino. In this way, the neutrino events are selected as the events with directions close to the sun [120] (Fig. 8.11).

The electron track must be sufficiently long to permit an adequate measurement; this limits the measurable neutrino flux to energies above 6 MeV. The number of observed events provides a measure of the neutrino flux.

The observed fluxes in the three detection methods, each sensitive to different portions of the solar neutrino spectrum, can then be compared with the fluxes expected in the solar model calculations. The results of all three observations, Cl^{37}: 0.36 ± 0.05, Ga^{71}: 0.52 ± 0.06, and Super-Kamiokande: 0.47 ± 0.09 (the error is dominated by the uncertainty of the solar model prediction), are below unity, indicating that some of the neutrinos have oscillated to other flavors on their way from their birth in the center of the sun to us.

Neutrino oscillation has also been observed by studying neutrinos produced by the cosmic rays in the earth atmosphere. High energy particles, mostly protons, with a wide spectrum of energies but typically a few GeV, circulate in our galaxy, and are incident on our atmosphere, at a rate of about 1,000 per square meter per second. In collision with the nitrogen and oxygen nuclei of the upper atmosphere, typically above 15,000 m, pions and kaons are produced, which in turn quickly decay, producing a flux of muons and muon neutrinos. In turn, the muons decay to produce muon and electron neutrinos, in equal numbers. These neutrinos can interact with the nuclei in the water of (Super-)Kamiokande, producing, respectively, secondary electrons or muons, which are detected and identified, and so the electron and muon neutrino fluxes are measured and also their energies roughly determined. The flavor identification can be appreciated from Fig. 8.12, which shows the patterns produced in the photo-multiplier array by the two types of particles. The electron "ring" is much fuzzier than that of the muon. The absorption of the neutrinos in traversing our planet is negligible. The direction of the neutrino is measured from the direction of its electron or muon, and the experiment consists of observing the rate as a function of its direction with respect to the normal to the earth at the position of the detector. Neutrinos which come from above have typically 15 km in which to oscillate, those from below have as much as 13,000 km. The results [121], Fig. 8.13, are very clear: Electron neutrinos are seen in the expected numbers, independent of whether they have come from above or below or in between, as expected if oscillation effects are negligible. This is not in conflict with the solar neutrino results, given the huge difference (factor of 100) in energy, as well as in distance available for oscillation. For the muon neutrinos, the observed flux coming from above is about as expected without oscillation, but the one from below is less by a factor close to 0.5, indicating strong oscillation, with oscillation length somewhere between the height of the atmosphere and the diameter of the earth, which, given their energies, corresponds to a difference in the squares of the masses comparable to 10^{-3} eV^2.

The observations of solar neutrinos and of the neutrinos produced by cosmic rays in the earth's atmosphere demonstrated that neutrinos oscillate from flavor to flavor. The solar data are best understood as $\nu_e - \nu_\mu$ oscillations, with mass difference $|m_{\nu\mu} - m_{\nu e}| = (.008 \pm .004)$ eV and mixing angle (45° corresponds to maximal mixing) of $(32 \pm 5)°$, and the atmospheric neutrino data as $\nu_\mu - \nu_\tau$ oscillations, with mass difference $|m_{\nu\tau} - m_{\nu\mu}| = (.05 \pm .02)$ eV and a maximum mixing angle, $(45 \pm 15)°$.

These results, which were recognized with the physics Nobel Prize of 2002, are convincing but nevertheless somewhat frustrating, because they all depend on missing neutrino fluxes; it would be reassuring to actually see the oscillation. Several experiments are in progress to check these results, some have already succeeded. Two of these study the solar neutrinos using new

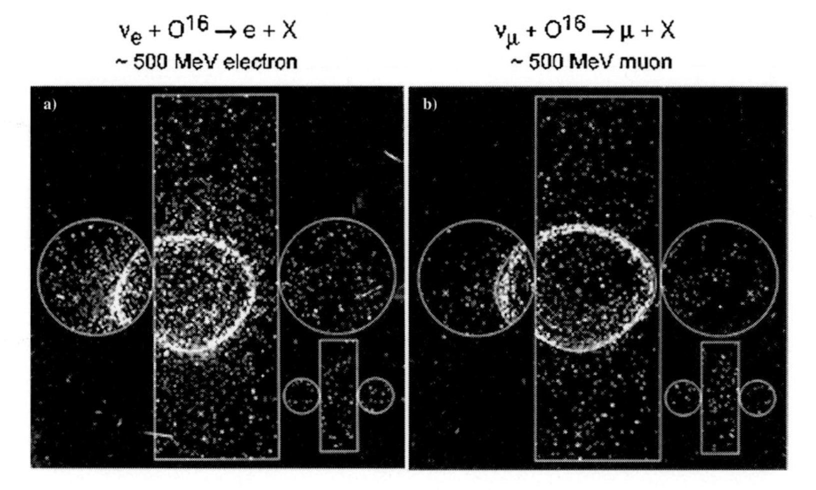

Fig. 8.12. Patterns produced by neutrino interactions in the Super-Kamiokande detector: **a**) a 500 MeV electron produced by an electron neutrino and **b**) a 500 MeV muon produced by a muon neutrino. In moving through the water with a velocity close to that of light, the particles (electron or muon) emit light in a cone around their direction of propagation. This is seen as the circle of photomultiplier tubes which produce signals. The electrons, in addition, degrade to produce secondary electrons and positrons at small angles, which in turn radiate light and produce the fuzziness. The difference in pattern permits the identification of the neutrino flavor, electron neutrino or muon neutrino

techniques, one in Italy, one in Canada. The latter has already produced an important result. It looks at the same part of the solar neutrino spectrum that was investigated by Kamiokande, but the vessel is filled with heavy water and, in addition to the interaction rate on electrons, it is able to measure the interaction rates on deuterons, which provides an important check. An experiment in Japan is measuring the electron neutrino flux from reactors, but now many reactors, at distances of hundreds of kilometers. The flight distance was for the first time sufficiently long so that in using a liquid scintillator detector in the old Kamiokande site, similar to but larger than that of Reines and Cowan [70] in which neutrinos were first detected. The disappearance of reactor electron neutrinos could be observed, and thereby the solar neutrino result was confirmed. Several so-called long baseline experiments are under construction in Japan, the U.S., and Europe to study neutrino oscillations. These experiments use accelerators, which produce muon neutrinos, and very massive detectors, typically tens of thousands of tons, at distances of hundreds of kilometers. A Japanese experiment, which uses the Super-Kamiokande detector and a proton accelerator some 250 km away, has produced a first result confirming the $\nu_\mu - \nu_\tau$ oscillations, and after a few more years of operation, on the basis of the measurement of the effect as a function of neutrino energy, it should be possible to really "see" the oscillation in the variation with neutrino energy

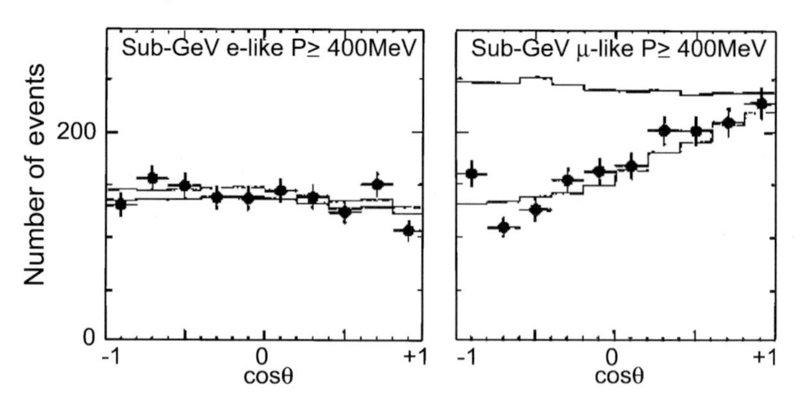

Fig. 8.13. Super-Kamiokande results for neutrinos produced by cosmic rays in the earth's atmosphere. The number of observed events is plotted as a function of the cosine of the angle of the observed electron (produced by an electron neutrino), left; and the observed muon (produced by a muon neutrino), right, with respect to the detector. $\cos\theta = 1$ corresponds to the top, the neutrino travelling only through the atmosphere, on the average about 15 km, and $\cos\theta = -1$ corresponds to the bottom; the neutrino has traversed the earth, about 13,000 km, and the other angles are in between. The expectations for no oscillation are the more solid, essentially horizontal lines. Electron neutrinos show no oscillation (the relevant mass difference squared divided by the energy is too small), but the muon neutrino flux traversing the earth is only one half, the other half has presumably oscillated to tau neutrinos

of the fraction of the neutrino flux that has oscillated away from the detected muon neutrino flavor.

Why is all this interesting? If one would like to understand particles, it is, of course, important to know the masses and mixing angles of the neutrinos, but there is in addition, and perhaps more interesting, the hope that the pattern of masses and mixings might give a clue in the search for the still missing fundamental theory of flavor symmetry.

So, in the domain of solar and atmospheric (cosmic ray–produced) neutrinos, in four different processes, effects have been observed that can only be understood as evidence of neutrino oscillations. The experimental results are more extensive than I described: for solar neutrinos the day-night variation has been measured (in the latter case, the neutrinos traverse the earth), and for the atmospheric experiments, the dependence on the neutrino energy could be measured. The extraction of the underlying neutrino mixing parameters, the neutrino masses and the mixing angles, is not trivial; it is complicated by the fact that in propagation through matter, such as the sun or the earth, electron neutrinos are affected differently from the other two flavors, a phenomenon which can give rise to dramatic mixing effects. The existing experimental results, solar, atmospheric, nuclear reactor, and accelerator (I am leaving out one experiment performed at Los Alamos National Lab which, if confirmed, would complicate the interpretation enormously) can be understood in the following way:

1) Electron and muon neutrinos oscillate from one to another with a large mixing angle and an oscillation length corresponding to a difference in the squares of their masses of between 10^{-4} and 10^{-5} eV^2.
2) Muon and tau neutrinos oscillate from one to another with a large mixing angle and an oscillation length corresponding to a difference in the squares of their masses of $(2.5 \pm 1) \times 10^{-3}$ eV^2.

In the frame of what we now know about particles, these experiments do not permit any other interpretation, but it would be reassuring to have some more direct evidence such as the production of a tau lepton by a neutrino known to have been created as a muon neutrino. Experiments in which neutrinos produced at accelerators are detected at distances of about 800 km are now under way or under construction both in the U.S.A. and Italy. The detectors are necessarily very massive, given the decrease of neutrino flux intensity with distance and the weak interaction of the neutrinos, but with a bit of luck we will have this confirmation in a few years. These experiments should also help us to complete the picture of the mixing "matrix." We now have a smell of four quantities: two differences in squares of masses and two mixing angles. But there are altogether seven numbers we would like to know: the three masses and four mixing angles for the three flavors of neutrinos.

Why is this interesting? The hope is that knowledge of this matrix of neutrino mixing parameters might help in finding the solution to the basic unsolved questions in particle physics. Despite the great advance in our understanding, there are still the great mysteries, as already mentioned: Why are the masses and interaction strengths what they are? It is generally imagined and there is some concrete evidence for the notion that at some much higher energy, at the so-called Grand Unification energy scale of 1,015 GeV (the highest laboratory energies at present are lower by 12 orders of magnitude!), the particles have no masses and are related by some fundamental, as yet unknown, symmetry pattern. The physics of particles at these high energies is also of great interest for the understanding of the evolution of the early universe. One can hope that the pattern of the neutrino masses and mixing angles, together with the corresponding pattern in the more accurately known masses and mixings of the quarks, might help someone develop the idea for the future theory of everything.

9

Experiments with the LEP e^+e^- Collider

The CERN SPS proton accelerator, which provided us with neutrinos, began functioning in 1976. This was the signal for the accelerator designers to think about a next project. CERN quickly focused on an electron–positron collider for two good reasons. In general, it was clear that the future of accelerators was with colliders, in which particles of equal energy, moving in opposite direction, are allowed to collide and interact. The energy which characterizes a collision is the energy in the center of mass (CM) of the system of the colliding particles. This only increases with the square root of the accelerator energy if the target particle is at rest, but it is just twice the accelerator energy for colliders. The first successful attempts at such collisions were made in 1961, at Frascati, Italy, famous for its wine. Electron-positron collisions with CM energy 0.4 GeV were achieved. A spectacularly successful e^+e^- collider was the 8 GeV CM Spear collider at Stanford University, which saw the discovery in 1973 of the new quark type (or flavor) called charm, completing the second quark family, and one year later, the discovery of the tau lepton, starting the third lepton family. Already in 1969 CERN had what was the first proton-proton collider, with CM energy of 60 GeV. In the early 1980s, it was possible with the imagination and inventiveness of Carlo Rubbia and Simon Van der Meer to operate the SPS as a proton-antiproton collider (also a first), with CM energy of 270 + 270 GeV, compared to the 28 GeV available when the 400 GeV protons of the normal SPS hit a fixed target. This SPS collider made it possible, in 1983, to make the very important discovery of the heavy, intermediate Bosons, the W^+,W^-, and Z, which had been postulated in the Electro-Weak theory to transmit the weak forces. The first reason to focus on LEP, the Large Electron-Positron collider, was a technical one: it was possible to build what would probably be the ultimate e^+e^- circular collider. The practically attainable energy in these e^+e^- colliders, in which the colliding beams are stored in rings with the help of magnetic fields, is limited by the energy loss due to the electromagnetic radiation of the stored particles, which for a given magnetic field increases with the fourth power of the energy, and so rapidly gets out of hand as the energy increases. A center of mass energy of about 200 GeV seemed a practical limit for electrons and positrons (this radiation problem is inversely proportional to the square of the particle mass and therefore negligible for proton colliders). The second reason was one of physics. At the time, the heavy weak bosons had been predicted but not yet discovered, and it was expected that they would be produced in LEP. As it turned out, and as we saw above, by the time LEP came into operation, these particles had already been discovered, but LEP turned out to be the ideal instrument to study their properties and to test this new theory with precision. At the time, Burton Richter, co-discoverer of the charmed quark at the e^+e^- collider at Stanford, put these arguments forward clearly. Although the final decision to build LEP, the Large Electron-positron Collider came only in 1982, already in 1980 LEP had become the focus for the CERN future, and some of us in the CDHS Neutrino experiment were joined by colleagues from France and Italy for a first discussion of how one might do an experiment at LEP. Perhaps we were 30 in this start, which evolved into a collaborative effort of some 400 physicists and lasted two decades. I entered this discussion with a clear wish: Could we think of some particular, clearly focused experiment which would require a special purpose rather than a general purpose detector? We could not, so we started to consider how we might build such a general purpose detector. We were not alone without an idea, no specialized detector was built for LEP or for any other high energy e^+e^- collider.

In the collisions of dominant interest, as the first step, the electron–positron pair annihilates; the energy is taken up by a "virtual" intermediate particle, which, in the Electro-Weak theory, is either a photon or a Z particle. This then decays into the particles which are finally observed. Since the virtual particle is produced at rest, this implies that the particles to decay will be emitted in all directions with comparable probability. Already in the previous generation of e^+e^- colliders, the detectors were designed to cover all outgoing directions as well as possible, with an inner part devoted to tracking the charged particles in a magnetic field in order to measure their directions and momenta. The tracker is surrounded by layers of "calorimeters" to measure the energies and directions of the neutral particles and, if possible, help identify the particle types. In such a detector, neutrinos are not detected at all, and muons, which in general do not produce showers in the calorimeters, are measured in the tracker and identified by their penetration. Particles can sometimes be distinguished by differences in their behavior in the calorimeters and by their ionization strengths in the tracker. All LEP detectors shared this basic conceptual design, but they differed in the techniques used in the implementation. Our design evolved rapidly. It was an interesting challenge, but it was also clear that it was possible to design a detector that would be adequate for the study of all the physics that we could expect from LEP, a situation quite different from that presented by the Large Hadron Collider, now under construction at CERN as the follow-up to LEP. The basic differences are two. In the first place, in the e^+e^- collisions in LEP, events were expected at the rate of the order of several per second, but given the strong interaction of the protons at the LHC, this is expected to be more than a million times higher. In the second place, in LEP the average event consisted of typically ten tracks, but at the LHC this will be more than 100, largely because the energy is 100 times greater, but also because our electrons are elementary particles, whereas hadrons are composite (sometimes referred to as "garbage sacks"). Despite significant progress in particle detection and data processing in the intervening two decades, the design of a clearly satisfactory detector for the LHC is not obvious.

We had open meetings about once a week, at which all important design features of what later was called the ALEPH detector were discussed and decided. We agreed to try to keep the design as simple and as uniform in technique as possible. The most important decisions were:

1. The magnet should be a superconducting solenoid with 1.5 tesla (15,000 gauss) magnetic field strength. Because of this high field strength and the required large size, 5.5 m in diameter, 7 m long, this represented a technical challenge.
2. The main tracking should be by means of a "time projection chamber" (TPC). The TPC is a very beautiful and conceptually simple device invented a few years earlier by David Nygren, an old friend and former collaborator, for the PEP e^+e^- collider at Stanford. It is a gas-filled cylinder in which the electrons, liberated from the gas atoms by the passing charged particles, are drifted to the ends of the cylinder by a strong electric field. There they are detected with the help of proportional wire chambers, and their positions and arrival times are measured. This permits a three-dimensional reconstruction of the track, with a precision of about 0.1 mm in the transverse dimensions and about 0.7 mm along the drift dimension.
3. The electromagnetic calorimeter should be optimized for spatial rather than energy resolution, in line with its important role of particle identification. This fundamental insight came to us from Jaques Lefrancois, a French Canadian and professor at Orsay near Paris. It should be inside the magnet in order not to suffer from the degradation of the particles by their interaction in the magnet coils and tank. The result was a 45-layer sandwich of 3-mm lead sheets separated by 5-mm-thick wire chambers, capacitatively coupled to square pads, about 3 cm $\times$ 3 cm, arranged in 75,000 towers, projecting to the collision point. The energy deposited along the way in the towers was sampled in three stories.
4. The hadron calorimetry should use the iron return yoke of the magnetic flux, which, conveniently, needed to be of similar thickness as that required for the hadron shower development. The iron was divided into 24 layers, each 5 cm thick, the layers were separated by 2 cm gaps, which accommodated the simplest and most economical detectors known, so-called "streamer" wire chambers, which sensed the number of traversing

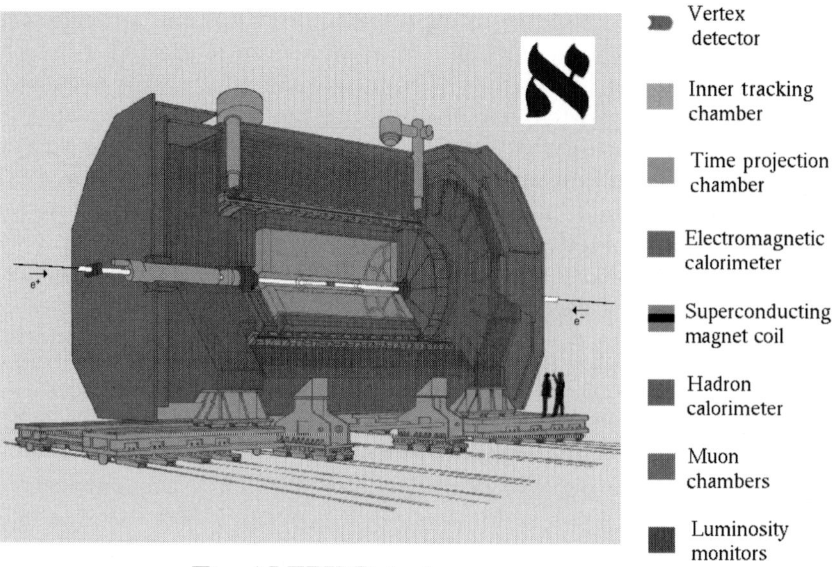

Fig. 9.1. View of the Aleph Detector

particles. These could be read out as wire planes and, capacitatively, in 6,500 projecting towers.

5. The detector naturally consisted of a central "barrel" and two "endcaps." It was agreed that the calorimeter technologies should be the same in barrel and endcaps, and that the two detection planes surrounding the whole detector, signaling the muons that had traversed the hadron calorimeter, would be of the same streamer chamber design as the hadron calorimeter detectors.

In retrospect, the design evolved remarkably rapidly. The detector that was proposed by 19 laboratories representing seven countries in January 1982 differs only in detail from the one shown in Fig. 9.1, which we ended up with seven and a half years later. By that time we were 32 laboratories. The construction cost, exclusive of laboratory personnel, was about 80 MSF, 20% of this from CERN, the rest assumed by the participating countries, according to their involvement and ability to pay. The experiment was approved by the LEP Committee in July '82.

Final design and construction proceeded in a spirit of collaboration and mutual pleasure. We were greatly aided by our technical coordinator, Pierre Lazeyras of the former CERN Track Chamber division, who anticipated and resolved technical problems and coordinated both construction and finances so that I, as "spokesman," never had to concern myself with such matters. Technically, one of the most challenging tasks was the superconducting solenoid. This was entirely and beautifully taken in hand by engineers of the Saclay

Fig. 9.2. The superconducting coil on the way from Paris to Geneva

laboratory, near Paris. Figure 9.2, shows the assembled coil on the road as it passed through a village close to Geneva, in 1987.

Looking at our proposal to track with a Time Projection Chamber, one of the LEP committee members, Prof. Guenter Wolf of Hamburg University, was not convinced that we would be able to achieve the accuracy we claimed (and needed), particularly because the only two previous attempts at TPC construction had not succeeded (for different reasons), and we had not addressed this question adequately in our proposal. Wolf came to me with his question, and already this was a special pleasure, since it was clear that it was a good question, and it is not often that a member of a committee understands the technicalities of a proposal as well as those who propose it. The problem was taken in hand by two young Italian collaborators, Francesco Ragusa and Luigi Rolandi. Within a few months, they had managed to understand the basic relevant physics, written a report, and could convince us and Wolf that the device would work as expected. Their work was confirmed two years later when, using a small-scale TPC model with a similar fundamental design, we could experimentally study the resolution as a function of the magnetic and electric fields. This was perhaps the most satisfying moment for me in my 15 years with this experiment. It would be easy to continue with other occasions of human satisfaction in this very international enterprise. I will mention one more, and that was the successful collaboration, between our Pisa and Beijing colleagues, in the construction of the more than 10,000

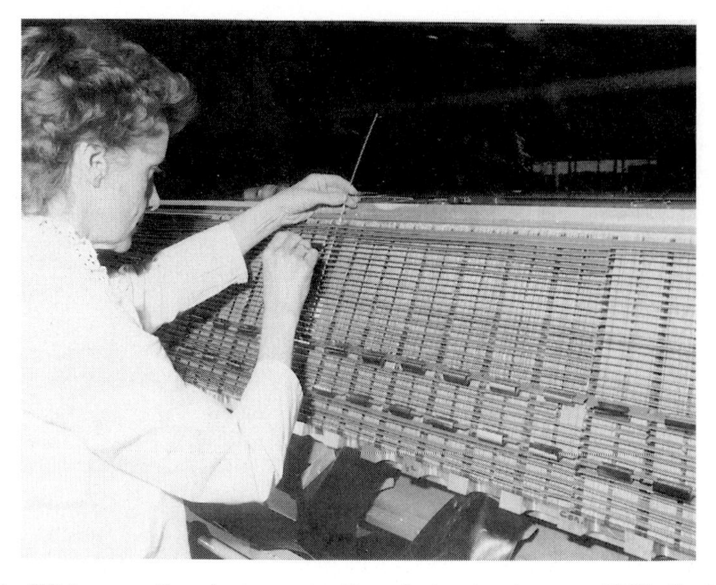

Fig. 9.3. Wiring up the electromagnetic calorimeter towers, 1986, Saclay (with permission by Jack Steinberger)

square meters of streamer chambers needed for the hadron calorimeter and muon detector. This type of detector had been invented some years before at the Frascati laboratory in Italy and had been selected for economical reasons. The Institute of High Energy Physics in Beijing, with no previous experience in detector construction, undertook our large production, in a happy collaboration with physicists from Pisa. It was a very large undertaking, especially by Chinese standards, and notwithstanding a substantial fire one day – fortunately without loss of life, but whose damage took a great effort to repair – all the needed chambers arrived in time for our experiment. Probably the most strenuous part of the detector construction was that of the electromagnetic calorimeter; the barrel in France, at the École Polytechnique, Orsay, and Saclay; and the endcaps in the U.K., at Rutherford Labs and the University of Glasgow. The final assembly required the soldering, by hand, of each of the 45 layers of the 75,000 projecting towers. Figure 9.3, shows an instant in this three-year effort. Figure 9.4 shows the assembled barrel, in 1988, and Fig. 9.2 shows the transport of the coil.

Aleph was much more complex than previous electronic detectors, with some 500,000 detection channels. Three-hundred thousand of these also required the measurement of the electronic pulse height, and 50,000, in the TPC, required very precise measurement of the arrival times. Altogether Aleph had roughly 15 times more electronic channels than our previous large experiment with neutrinos. A consequent worry for me was how to find the necessary computational facilities to deal with the large amounts of data that would be produced. As it turned out, I need not have worried. In the intervening

Fig. 9.4. November 1988, the assembled Aleph barrel, with left to right: Jacques Lefrancois of Orsay, myself, and Pierre Lazeyras of CERN

time, the speed and capacity of commercial computers increased even more, in particular with the introduction of work stations, and computer capacity never was a problem. On the other hand, reading out the electronics and recording the data in the computer turned out to be a much larger problem than we had been conscious of; we barely managed to resolve this in time for the start-up of LEP in the fall of 1989. The LEP collider operated until the end of the year 2000. The first six years, LEP 1, were with CM energies near the Z "resonance" at 91 GeV. After that, LEP 2 operated at higher energies, up to a maximum energy of 209 GeV. This can be seen in Fig. 9.6, in which the e^+e^- collision probability is plotted versus CM energy, both as experimentally observed and theoretically predicted in the Standard Model. As can be seen, the resonance boosts the reaction rate by more than a factor of 100. The energy of the peak is a measure of the Z mass, not very precisely known

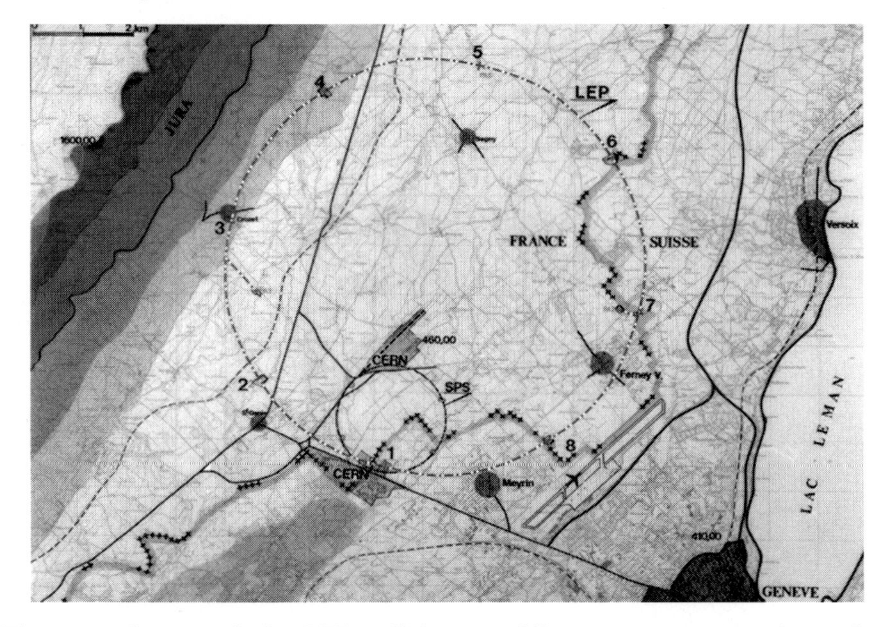

Fig. 9.5. Layout of the LEP collider, straddling two countries: the 28-km-circumference LEP ring, which was injected with 20 GeV electrons and positrons by the SPS ring, which in turn was fed by 2 GeV electrons and positrons by the PS ring, built in 1957, which in turn was fed by a newly built accelerator, producing 600 MeV electrons and positrons (© CERN)

before LEP, and the width (energy breadth) is a measure of its lifetime (= Planck constant/energy width), about 10^{-24} seconds, not very long!

The events observed at the resonance are the decay products of the Z into either a charged lepton pair or a quark-antiquark pair for each of the five lower mass quarks; there is not enough energy for the top quark (the top quark mass is 175 GeV). The quarks materialize as jets of hadrons; the five flavors cannot, in general, be distinguished, only if there is enough information on the particular hadrons that appear. The Z decays into neutrino and antineutrino pairs about 20% of the time. These decays are not detected, since the neutrinos don't leave tracks in the detector. Examples of the four main types of observed events are shown in Fig. 9.7.

Although the Z decays into neutrinos are not observed, the earliest, and one of the most important, results of LEP involved these particles. The width of the decay of a particle is just given by the sum of all the decay probabilities, and because of this, also by the height of the peak (the wider the peak, the lower it is). At the time, elements of three fermion families were known, all with very different masses. Would there be more families, at masses beyond the energies of the accelerators of the time? The neutrinos were known to have tiny masses, if any (they are now known to have masses, but less than one millionth of that of the next lightest nonzero mass particle, the electron).

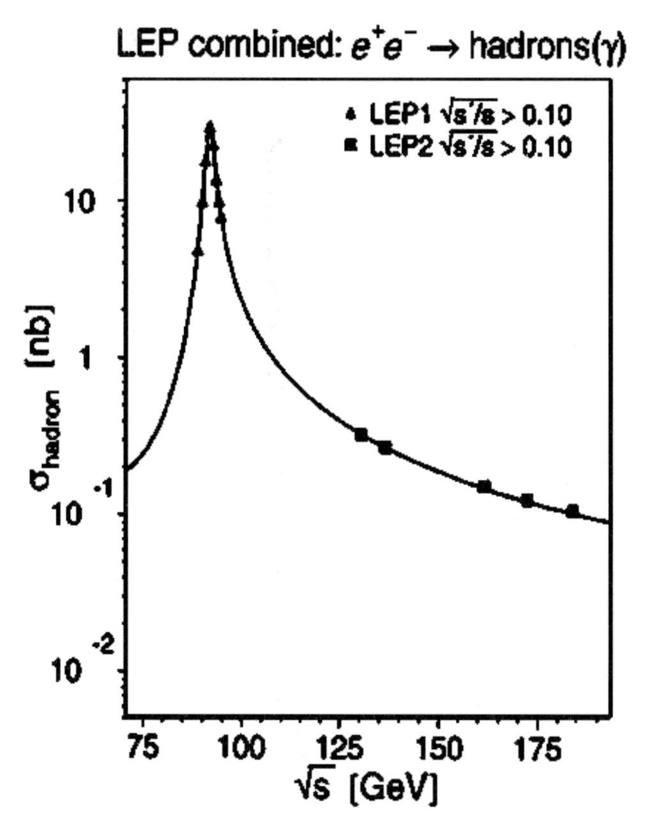

Fig. 9.6. Collision probability for hadronic final states as a function of the e^+e^- CM energy. The points near the Z resonance are with LEP 1, at the higher energies with LEP 2. The lines are the prediction of the Standard Model, given the LEP results for the Z^0 mass and width

It is natural to suppose that if there are higher mass families, their neutrinos would still be very light compared to the Z, the Z would therefore also decay into these neutrinos, and this would be duly reflected in a correspondingly larger Z width and lower peak amplitude. What was seen, however, in the fall of 1989, during the first few weeks of LEP operation, was a height and width of the peak, which, in the frame of the electro-weak theory, corresponded to just three neutrino families. Our first Aleph result, after a few weeks of LEP running in October 1989, was $N = 3.27 \pm 0.30$. In November, the result had improved to 3.01 ± 0.16, and at the end of LEP 1, in 1995, the combined LEP result was $N = 2.990 \pm 0.015$ (Fig. 9.8).

In the beginning, our Aleph results were substantially more precise than those of the other three LEP experiments, because we had realized that, statistically, the more precise result could be achieved using the height of the peak rather than its width. But this required a corresponding understanding

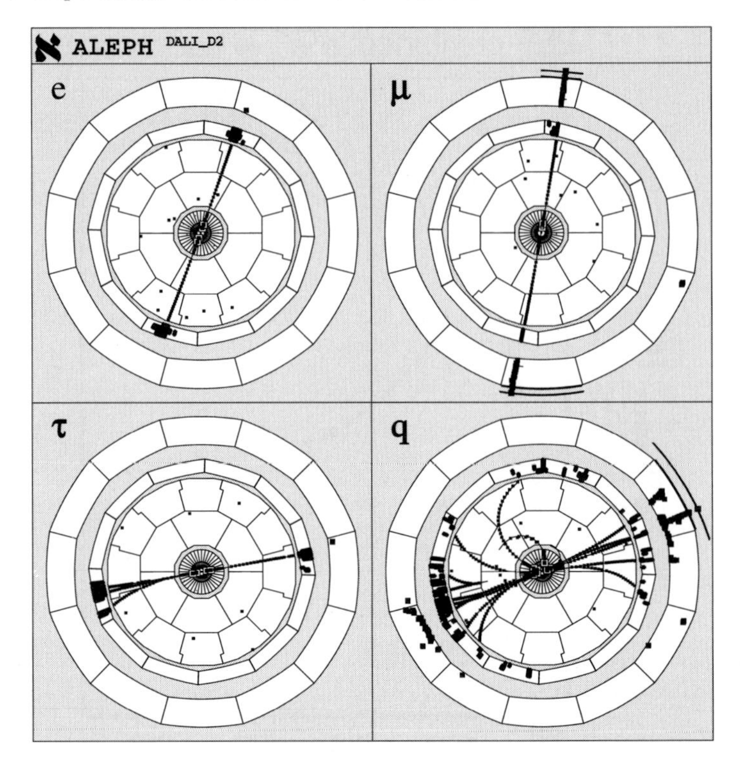

Fig. 9.7. Examples of typical events at the Z resonance. Top left, electron–positron. Each has one half of the energy and identifies itself by the shower it produces in the electro-magnetic calorimeter. Top right, muon pair. The muons identify themselves in penetrating the two calorimeters without interactions and activating the outer layers of muon detectors. Bottom left, tau lepton pair. The taus decay after typically 1 mm of flight, too short to be visible in the picture, although the decay can often be reconstructed with the help of the silicon micro-strip vertex detector. The taus can decay in several ways, sometimes with one, sometimes with three secondary charged particles, always with an undetected tau neutrino. Bottom right, decay of a Z to two quarks, which appear as jets of hadrons

of the "luminosity," that is, of the effective collisional flux, and we made a consequent effort to understand this luminosity more precisely. The fact that there are just three fermion particle families is one of the most important results obtained at LEP. The interactions of these particles are well described by the Standard Model, but why the number of families is three is still a challenging mystery, as are the other free parameters in the theory such as the particle masses and the three interaction strengths.

Some five million events of Z decay were accumulated during LEP 1 in Aleph; their analysis permitted substantial progress in several directions: precision measurements of parameters of the Standard Model and checks on its correctness, studies of the properties of known particles, and searches for new

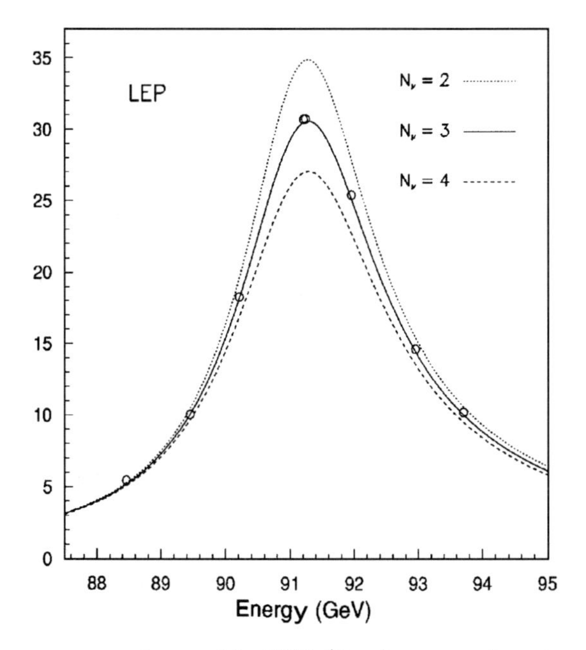

Fig. 9.8. Z resonance as observed in LEP (the four experiments combined). The circles are the measurements, the curve is the expectation of the Standard Model with three light neutrino families

particles. The first of these, the testing of the Standard Model, was perhaps the most important of these. Heretofore, its new predictions had been demonstrated qualitatively such as the existence of neutral currents and of the heavy intermediate vector bosons, but the measurements of their properties had been of very limited precision. At LEP 1 the interesting studies were: the precise shape of the resonance curve, the branching ratios to the different fermion channels, the angular distribution of these with respect to the direction of the incident electron and positron. In particular, this permitted precise measurements of the Z mass, the so-called Weak, or Weinberg angle, which describes the relationship of the neutral to the charged weak currents, and that of the W and Z masses. The Weak Angle could be related to several independent precision measurements, and their concordance was an important support for the theory. The improvement in precision achieved at LEP can be seen in that of the Weinberg angle. Before, it had been known with a precision of 10 to 15 %. At LEP, it was measured with a precision of better than one part in one-thousand. The most important confirmation of the theory was the precise verification of a subtle feature of the theory, the so-called radiative corrections. These affect the theoretical expectations of the measurable quantities at the one-percent level. Quantitative verification of these radiative corrections was a very important confirmation of the validity of the Standard Model. A by-product of this was a first "measurement" of the mass of the top

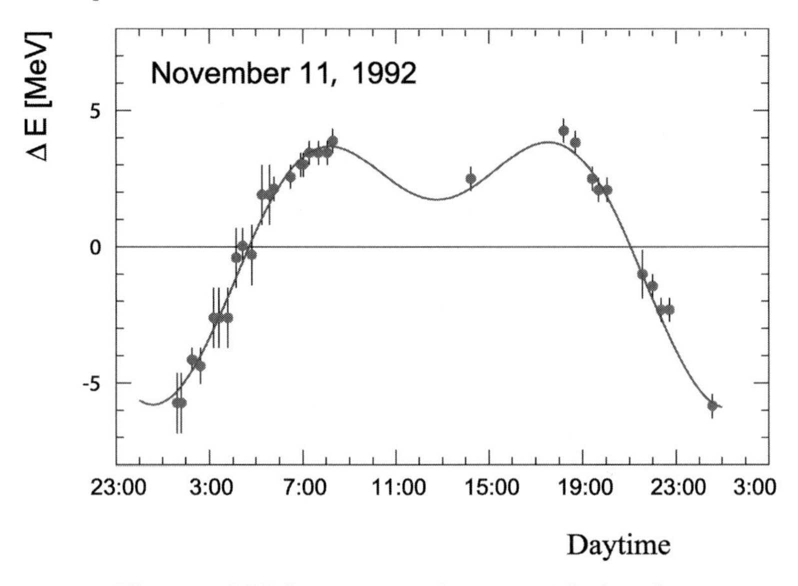

Fig. 9.9. LEP beam energy changing with the tides

quark, before it was actually discovered at Fermilab in 1995. Although the top quark is too heavy to be produced physically in Z decay, it enters "virtually" in the radiative corrections, and by comparing these theoretical predictions with the measurements, it was possible to deduce the top quark mass to be about 160 GeV, quite close to the mass of 175 GeV at which it was found a few years later at the Fermilab p-p collider.

These measurements required a precise knowledge of the beam energies. New, ingenious techniques for the measurement of the beam energies evolved, using laser light, which could sense the polarization of the resonant frequency of the electrons (or positrons). At a certain stage in the development of this technique, strange daily variations in beam energy were noticed, at a level of one part in ten-thousand of the energy. These were mysterious until a young American colleague, Bob Jacobson, one-time ocean sailor, another time president of a computing enterprise, now post-doc in physics, had the idea that these in fact were due to the moon tidal distortions of the earth's surface, the same gravitational forces that produce the ocean tides. An example of later, even more precise measurements is shown in Fig. 9.9. By this time, it is quite obvious. No one had thought of this effect before it had been observed. For me it was a delightful happening illustrating, on the one hand, the high technical quality of the work at LEP and, on the other hand, how science advances.

We have discussed the important contribution of LEP 1 results to the confirmation of the Electro-Weak theory. Although the process of Z production and decay is dominated by the weak interaction, the strong interaction contributes also, in its own radiative corrections, predicted in Quantum Chromo Dynamics, the theory of the strong interaction. These could be checked ex-

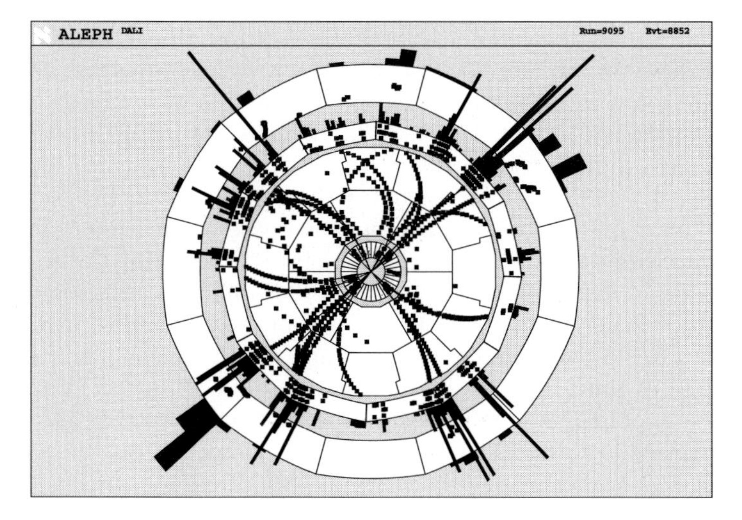

Fig. 9.10. Z decay into quark, anti-quark, and two gluons. All four particles manifest themselves as jets of hadrons

perimentally, providing important support to this new theory. A beautiful confirmation of the underlying QCD symmetry structure was provided by the relatively scarce events in which the Z decays to quark, antiquark, and two energetic gluons. Such an event is shown in Fig. 9.10. All four outgoing particles manifest themselves as hadron jets, basically indistinguishable, but the gluons are usually the less energetic of the four. The symmetry structure of the QCD theory could be confirmed on the basis of the observed angular correlations between the four outgoing particles.

It was also possible to make some progress in our knowledge of the properties of some of the newly discovered particles. The hadronic decays of the Z provided a better and larger sample of some hadrons containing heavy quarks than available before. Two new hadronic particles were found, a new meson composed of a b (anti-b) and an anti-c (c) quark, as well as a hyperon (nucleon with one or more quarks replaced by heavier flavor quarks) composed of down, strange, and bottom quarks. In this work, an essential element in the detector was the silicon chip vertex chamber, which permitted the tracking of charged particles in an inner volume about 25 cm in diameter and 30 cm long, with precision of about .01 mm, about 15 times more precise than the wire chamber tracking. This permitted the reconstruction of the decay vertices of b and c mesons, typically with decay lengths of the order of 1 mm.

One of the most intensive efforts was the unfortunately unsuccessful search for the "Higgs boson." The Higgs is a particle, a construct of the electro-weak theory, essential to its self-consistency. Without this Higgs, all the particles of the theory would be massless, which, experimentally, is not the case. The Higgs is a neutral particle of spin zero; until now, no particles of spin zero are known. Its mass is not specified by the theory, other than that it should not be greater than of the order of 1,000 GeV or else things go wrong. Its interactions, however, are specified in the theory, so its production probability in electron–positron collisions is a predicted function of its mass. No sign of the Higgs was found in LEP 1; a lower limit of 76 GeV on its mass (about 80 proton masses) could be put. LEP operation at the Z resonance energies terminated in early 1995, in order to convert LEP 1 to LEP 2, that is, to higher energies. This required a very substantial increase in the power of the radio frequency acceleration system, which was solved by replacing the copper resonance cavities by superconducting cavities. This difficult technology had been researched for many years at CERN for this purpose. The installation of the new cavities was a considerable undertaking, also financially, and proceeded over a period of time, with gradual increase in the CM energy from 150 GeV at the start-up of LEP 2, late in 1995, to the final energy of 214 GeV, at LEP 2 closure at the end of 2000. The higher energies offered the possibilities of particle searches at higher masses, both for the Higgs and for particles proposed in theoretical extensions of the Standard Model, as well as new checks on the validity of Standard Model predictions. All results for production rates and angular distributions were in agreement with the predictions, in particular, also those for the new channels opened at the higher energy in which $W^+ W^-$ pairs of heavy bosons are produced (Fig. 9.11).

An intensive effort was concentrated on the precision measurement of the W mass, because it is directly related, in the Standard Model, to the now very accurately known Weinberg angle. The comparison of the measured mass $M_W = (80.451 \pm 0.033)$ GeV with the predicted value $M_W = (80.379 \pm 0.033)$ GeV is at present the most sensitive check on the predictions of the E-W theory. The new particle searches were for the Higgs, as well as for particles in the extension of the Standard Model known under the name of Super-symmetry, or Susy. Some such extension of the SM in going to higher energies is necessary. Susy predicts a new set of particles of higher masses, as many new ones as we have old ones. Indirect evidence exists for some form of Susy. One of these comes from particle physics, where there are extrapolations that indicate that at very high energies, the E-W and the QCD partners of the SM coalesce into a unified theory if the extrapolation is carried out in the frame of Susy. The other comes from cosmology, where it turns out that it is possible to understand the amount of the invisible, non-barionic dark matter, which dominates the matter content of our universe, if it is supposed that this dark matter consists of Susy particles. Unfortunately, to the disappointment of our eager searchers, no new particles were found. A new, lower limit, 114 GeV, could be established for the Higgs mass, and this is a useful result. The failure to find Susy particles limits the values of the parameters in the several Susy models, but does not exclude the possibility that some form of Susy will turn out to be there. There is good reason to hope that both the Higgs and the Susy cosmic dark matter will be found at the Large Hadron Collider (LHC), now under construction at CERN.

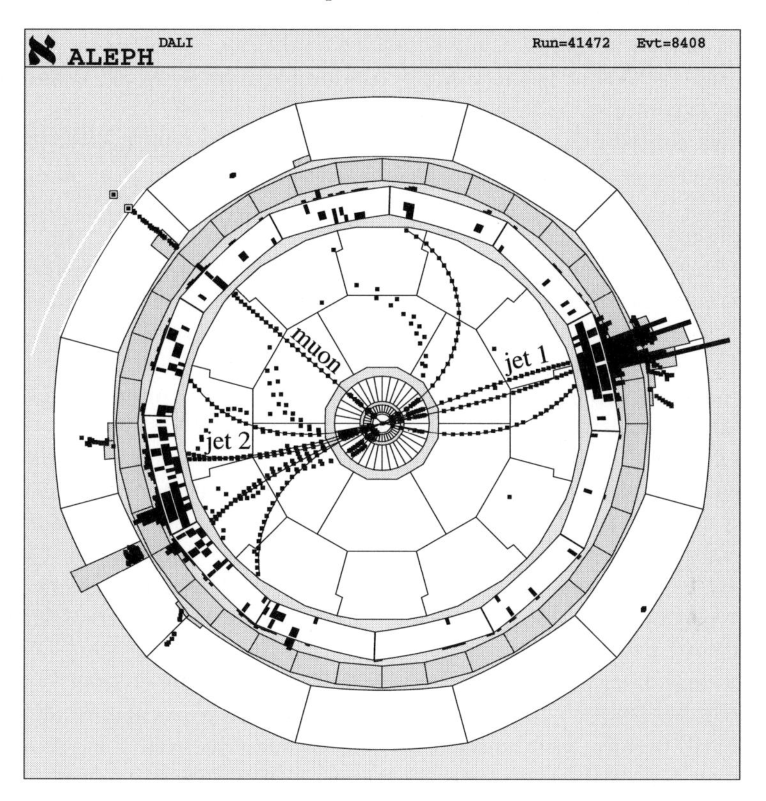

Fig. 9.11. Example of the production of a pair of W heavy bosons. The Ws decay in $\sim 10^{-24}$ seconds, essentially at the point of collision. The event shown is an example with collision energy only slightly above twice the W mass so that the bosons are produced with small velocity, almost at rest. One of the Ws decays into a quark and an antiquark, which appear as back-to-back (because the decaying particle is nearly at rest) hadron jets, the other decays to a muon and neutrino. The muon is the single track that penetrates the detector, the neutrino, of course, is invisible

Summing up, at LEP we have learned that there are three, and no more than three, fermion families; the electro-weak theory has been checked and confirmed, including its predicted radiative corrections, with precisions at a fraction of one part per thousand; precise values were obtained for several important parameters such as the Z and W masses and widths and the weak mixing angle; certain aspects of Quantum Chromo Dynamic could be confirmed and its interaction strength measured more precisely; several new heavy hadron states, mesons and hyperons, were discovered; and some useful lower bounds for the Higgs meson and possible super-symmetric particles could be established. It was an effort that extended over 20 years, required the dedicated commitment of 1,500 to 2,000 physicists, and a comparable number of technical and supporting personnel, at a cost of perhaps 10 to 20 billions of dollars (it is not obvious how to count this) paid for by our society. Was it worth it? What is the value of fundamental knowledge? What else did humanity accomplish in that period, and at what price?

My own contributions to the ALEPH effort, and for that matter to particle physics altogether, ended in 1995. In 1990, I ended my term as spokesman, but continued in the work of trying to learn what we could from the accumulating data. The last of this was the measurement of the branching fraction of the Z boson to b quarks, some two to three years of work which succeeded in clearing up what had been for some time a problem. Previous results had disagreed with the prediction of the electro-weak theory, but these results turned out to be in error. So came to an end almost 50 years of participation and pleasure in the quest for progress in our knowledge of the fundamental constituents of all matter, souls included. In this period, the particle physics scene had changed a lot. Cosmic rays were replaced by accelerators and accelerators were overtaken by colliders. A great deal has been learned about particles, and a beautiful theory of their interactions has emerged. So much has been learned that further progress has become enormously more difficult. In 1948, with my thesis, I could make an important step forward, alone, in a few months. Now, for the LHC, the proton collider that is being built in the same tunnel that housed LEP, but with an energy almost 100 times greater, some 5,000 physicists are engaged in building the four detectors. The collider is expected to function in the year 2007, and, hopefully, those who prepared all this and who will still be alive will be rewarded with some new basic insights. Perhaps it will then be 2010, more than 20 years after the program was launched. Progress in particle physics has become very difficult, but there is no lack of interest, because very important fundamental questions remain to be answered such as: Why are the parameters of the Standard Model, the particle masses and coupling strengths, what they are, and what is the nature of the overall symmetry that will combine the electro-weak and the QCD theories into one, "grand-unified" theory (GUT) at an energy scale of about 10^{15} GeV, a factor of 10^{11} higher than the LHC? What is the physics underlying the inflationary epoch of our universe, when the energy density was perhaps near the GUT scale?

Since I can no longer contribute to particle physics, I am spending my time, and enjoying, trying to learn about what is known and what is being learned in cosmology and the astrophysical observations on which it is based. It is exciting for me for two reasons. On the one hand, there is great progress, driven by advances in the observational techniques such as optical and X-ray satellite telescopes, radio telescope interferometers, and cosmic microwave background observation instruments. On the other hand, the basic physics which underlies the advances in our understanding of cosmology is very varied and often delightfully challenging.

As an example, I will try to give a glimpse into the progress of the last three years through observations of the cosmic microwave background radiation, the CMBR. In early 2000, the team of Boomerang, the balloon born

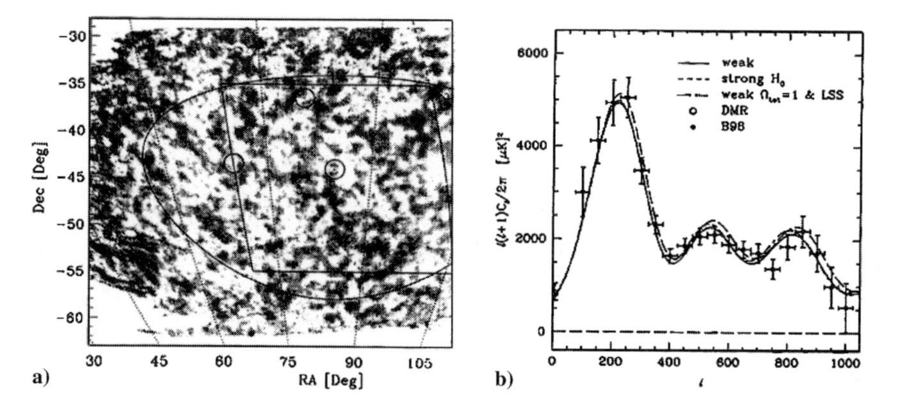

Fig. 9.12. **a**) Map of the temperature anisotropies measured by the Boomerang balloon borne microwave detector, and **b**) the "Power Spectrum" (square of temperature deviation from average) as function of the inverse angular scale l (l $\sim 180^\circ$/ angular scale in degrees) deduced from the map. The positions and magnitudes of the peaks and valleys permit the evaluation of several very important cosmological parameters

CMBR telescope, published the results of a lucky flight that circled the South Pole and came down just where it had been launched ten days before. The CMBR photons observed by Boomerang, at wavelengths not too different from those used in satellite TV communications, were emitted with wavelenghs 1,100 times shorter, and therefor with energies 1,100 times greater (the reduction in energy is proportional to the expansion in the linear dimension of our universe), at the time of "decoupling," when the ionized electron-proton plasma cooled sufficiently to form neutral atoms. Since then, the photons have travelled freely, without scattering, and so image the universe at that time. Boomerang measured the pattern of the temperature of these photons (the photons have, then and now, an energy distribution called a black body, or Planck, energy spectrum characterized by a temperature) over a portion of the sky, about 4% of the sky. Basically, the temperature is very uniform, but at a level of about one part in 100,000 there are variations. It is these tiny anisotropies that are interesting here, and the observational challenge is due to their smallness. The Boomerang sky map is shown in Fig. 9.12a, and the corresponding "power," that is, the square of the deviation from the mean, as a function of the angular scale of the deviation, averaged over the map is shown in Fig. 9.12b, where the Boomerang data are combined with several other more recent observations. The parameter l, which labels the abscissa, is 180° divided by the angular scale of the inhomogeneity. The bigger the l the smaller the angular scale, $l = 200$ corresponds to about a one degree angular scale.

The CMBR anisotropies were discovered in 1992 by the Cobe satellite, a most exciting event. Cobe had measured these at relatively large angular scales, $l < \sim 15$. Following Cobe, ground-based telescope gave some earlier indication of the important structure at higher l. Now, we clearly see several peaks and valleys in Fig. 9.12b. The general pattern was not unexpected, it had been predicted. The peaks and valleys reflect oscillations in the pressure of the gas when the universe, at the time of decoupling, was about 200,000 years old (now it is about 14 billion years old) and can tell us about the constitution of this gas and about the propagation of the photons since that time. It is the most detailed view of the early universe that we now have and permits very important new cosmological insights. For instance, from the position of the peak at $l \sim 200$ it can be deduced, with a precision of about 5%, that our universe is "flat." The matter of "flatness" or a possible "curvature" of the universe is one of the most basic properties of the Einstein-Friedmann equation, which governs the dynamics of our universe, and this is the first direct evidence we have of the fact that our universe is flat. And this is not the only striking, new knowledge that comes from just this observed scale dependence of the CMBR temperature anisotropies. One of the most interesting periods of modern cosmology is the very early split-second of "inflation," when our present universe is supposed to have expanded from about 10^{-20} cm to a few cm. This inflation period is essential to our understanding of the uniformity of the observed universe on large scales, as seen, for instance, in the CMBR, which is uniform, except for the tiny inhomogeneities at the 10^{-5} level. One of the achievements of inflation theories is that they predict not only this uniformity, but also the inhomogeneities observed by Boomerang. These, under gravity, have then grown into the structures we now see: the stars, galaxies, and clusters of galaxies. Without the early, small inhomogeneities, there could not be any structure, ourselves included. In the inflationary theory, these originate as a quantum mechanical effect, zero point fluctuations. The observations give a measure of the magnitude and the scale dependence of these primordial inhomogeneities, and their observation is a very important clue to our understanding of this earliest epoch. A good part of my time for a year or two has been devoted to trying to learn some of the basic physics essential to the understanding of the CMBR observations. Much of this physics I had escaped learning, as I was focusing narrowly on particle physics: the quantum mechanics of the primary fluctuations and their conversion to classical objects (I am afraid that I haven't yet succeeded in understanding the latter), the general relativity necessary to understand the evolution of the inhomogeneities while they are outside the "horizon" (when their scale is larger than the Hubble distance) and after they re-enter, and the hydrodynamics of the acoustic oscillations before decoupling. Trying to learn some of this basic physics has been fascinating, even if sometimes frustrating, given the limitations of old brains.

I have tried to give an account of my 50 years in particle physics and some idea of the basic advances in this field during this period. The beautiful present understanding of the physics of particles and their interactions is, in my opinion, one of the cultural achievements of the century just gone by, and my opportunity to participate in this was a privilege. We have learned a great deal. We know that there are three fermion families and all of their members are known. There are three interactions, transmitted by three different sets of vector bosons, and the beautiful gauge theories of these interactions have been tested experimentally, with high precision. But the present generation of young physicists can be content: We have left you many interesting questions. Why are there three families and how are they related? Why are the masses and the mixing parameters what they are? There are clear hints that at energies ten orders of magnitude higher than those which have so far been reached in the laboratory, all three interactions become unified. What is the theory and what are the symmetries of this so-called Grand Unification regime? What is the "dark matter" that dominates (or dominated) the energy content of our universe? What are the particles that underlie the "inflationary" era and the recent acceleration of the expansion of our universe? How can a quantum theory of gravity be constructed? There is no lack of unanswered, fundamental questions. But the remaining questions are more difficult than those clarified by my generation.

A

Postlude

Family

Let me finish this biography with some more personal comments. I have two families. My first girlfriend, who a month or so later became my first wife, and I met in the fall of 1942. Joan, from a family of French origins who had settled in northern Wisconsin, was secretary in the office of the U.S. Army Signal Corps in Chicago, the office organizing the courses in basic physics I was attending at the University of Chicago as an off duty enlisted soldier. It was a civil marriage; the only present we received was a bar of chocolate. Some months later, to appease Joan's family, following instructions by a Catholic priest, the marriage was redone by the church. Joseph Ludwig was born in November 1944, while I was assigned to the Radiation Laboratory at MIT, developing radar bomb-sights. An incident that gives some insight into my shortcomings, and of which Joan reminded me much later, is that when Joan came into labor, I took her to the hospital in Boston and went back to my work, without waiting for the arrival of the offspring. In retrospect, it is clear that I am to blame for not devoting enough time to my children. Some six months after Joe came, in the summer of 1945, shortly after the surrender of Germany, I was called to active duty in the army, and Joan was alone with Joe until my discharge, February 1946, when we found each other again in one of the small prefabs that had been erected for returning soldiers at the University of Chicago, south of the Midway. In November of 1948, Richard Ned later famous for the design of electronic musical instruments, was born in Princeton, where in September, at the Institute for Advanced Study, I had started a postdoctoral job. I hope, but don't remember, that this time I was present for the arrival of the newborn. In the mid-fifties, after we had settled down at Columbia University, Joan began to take a serious interest in painting. We lived in Hastings-on-Hudson, a well-off New York suburb with good schools for the two boys. Joan attended art classes at the Art Students League, in midtown New York, and acquired a new circle of friends. The marriage began to deteriorate, I was accused of intimacies with other ladies.

End of 1960, our wife and mother left Hastings for a room and attic space to paint, on New York's Lower East Side.

A Mexican divorce, in the summer of 1962, was followed in November by marriage to Cynthia Alff, my very pretty and extraordinarily gifted Ph.D. student at Columbia. Cynthia had grown up in the Queens section of New York City. Cynthia's mother came from a family of Swedish immigrants, her father, whom Cynthia never knew – he had died when she was only some months old – was the son of German immigrants. To put things into some perspective in time, the experiment in which the second neutrino family was found began in 1961 and ended during the summer of 1962. The wedding was not more elaborate than the first, but now we had three witnesses, our two mothers and Ned, Joe was too busy at a football game. I moved out of the house in Hastings, Joan moved back in, and Ned, now 14, stayed with Joan. Joe had started his college studies at Columbia and was quartered there. Luckily, both were old and mature enough so that they could take this calamity in stride, without, to the best of my knowledge, suffering a great deal. Cynthia and I found quarters, a floor in a house of friends, still in Hastings, a beautiful early 19th-century mansion overlooking the Hudson River.

And so the second family began. Cynthia finished her thesis in 1963, an experiment at Brookhaven National Lab, in our 30″ hydrogen bubble chamber, which showed that the parity of the Λ and Σ hyperons are the same, as was expected in the model that had been proposed by Gell-Mann and Zweig, the model that first put forward the idea of "quarks." I sat on the Ph.D. committee of my wife, perhaps a questionable procedure. Cynthia passed. In 1964, I could take a sabbatical. We chose to split this, one semester at the recently founded University of California in La Jolla near San Diego, in southern California, the second at CERN, where I had already spent some summers. The beautiful drive from New York to La Jolla, via the North Dakota Bad Lands National Park, some hikes in the Colorado mountains, the Grand Canyon north edge, including an overnight hike to the bottom, Bryce Canyon National Park, and some days in Yellowstone served as our belated honeymoon trip. In La Jolla, Cynthia found a job with Berndt Mathias, growing crystals to serve in the study of superconductivity. I gave a course in particle physics and thought a bit about a rather basic experiment to elucidate CP violation, which had just been discovered in the spring. They were a good few months; we found interesting friends, both among the physicists at the university and among biologists at the Salk Institute nearby. We enjoyed an excursion to the nearby Baja Peninsula, in Mexico, with a very different culture. There was some mountaineering in the Sierras and a kayak trip on a desert lake. The most absorbing happening for me in La Jolla was the surfboarding in the Pacific a few meters from the university every day at noon, but this I have already mentioned.

Returning from CERN to New York in the summer of 1966, Cynthia switched from physics to biology, taking a post-doc job in molecular biology at the New York University School of Medicine, under Prof. Werner Maas, studying bacterial genetics. When we moved to Geneva in January of 1968, Cynthia continued this at the University of Geneva with Professor Kellenberger, before beginning her job at the World Health Organization. I believe that it is not without interest to remember here that Cynthia was delayed in starting the WHO job by an agreement between the United Nations and the U.S. government, which required (perhaps it still does) the UN to get U.S. security approval, conditional to the employment of U.S. citizens. As we learned only a few years ago, on reading relevant old FBI records made available under the U.S. "Freedom of Information Act," Cynthia was under suspicion because of her association with me. The WHO job was not particularly satisfying, and Cynthia could appreciate several of her colleagues, in particular, the Russian director of the group, until it came to an end in the wake of a WHO re-management.

In the meantime, we had moved into the house which had been built for us, while we were still in New York, in nearby France, in the style of a local farmhouse, beautifully located in the fields on steeply sloping grounds at the foot of the Jura mountains, with a view over Geneva, the Alps, and Mont Blanc. Our first child, my first daughter, Julia Karen, arrived on March 24, 1974. We had waited a bit, first trying to convince ourselves that our marriage was adequately stable. I was 53, Cynthia 35. Cynthia continued to work, half-time now at CERN, back to physics. Julia's brother, John Paul, appeared on May 19, 1977. When the time had come for Julia to begin school, we were a bit afraid that the French school system might be very rigid, and in 1981 moved to a house we had built in Geneva. The new location and the new house were less beautiful, but the Geneva schools were, in fact, excellent; I have always been most grateful to Geneva for its good schools. The move also made it possible for Cynthia to work in her preferred subject, biology, since we were now rather close to the University of Geneva, where Cynthia found a job as chargé de recherche in molecular biology. Both Julia and John were able to do their "maturité" in the Geneva "collège" (high school) system, and then continued to do undergraduate studies at Brown University, Providence RI, U.S.A. At this writing, Julia is trying to finish her Ph.D. in low temperature physics at MIT, before dropping physics for politics, and John has completed his Masters in mathematics at Waterloo University in Canada, and he is continuing his studies at the University of California–Davis.

Joe, now 58, after teaching school in Canada and studying at a theological seminary to avoid participating in the Vietnam War, finished his law studies at Columbia University. He and his girlfriend Ellen then went to Maine, one of the original British colonies on the new continent, and there, in the virgin woods, hand built their tiny wooden cabin, without plumbing, an outhouse for toilet, managed to survive several years, and Joe received accreditation to practice as a lawyer in Maine. But law was never more than one of Joe's occupations. After some years in the log cabin, he moved to Augusta, capital of Maine, to work for Common Cause, a U.S. national group engaged in the promotion of the interests of the consumer and the lower castes. He achieved state-wide recognition in leading a legal challenge by Common Cause of some questionable undertaking of the Bath Iron Works, constructors of U.S. Navy vessels. From there, he moved to Rockland, Maine, a town of 15,000, where he has been since. He set up a private law practice, and Julie Schultz and he were married. Together, they built their own modest but comfortable wooden house. Julie, together with Joe, founded the Penobscott School (named after Penobscott Bay, on which Rockland's harbor lies), which teaches languages, in the school season foreign languages to locals, and in the summer English to foreigners. At the same time, it contributes in a beautiful way to the local social culture, in that the summer students (adults) are housed as guests by the local citizenry, and so promotes international understanding. One of Joe's pleasures is the repair of the old wooden houses that are the tradition of Maine. He writes personal op-ed pieces for the local newspaper, on cultural and political issues. In the last years, he has served on the Rockland City Council. His keenest interest these past few years has been the local public radio station he created, and which has received national recognition in *The Nation* magazine.

Ned had remarkable little gift for academics. He did not complete his first semester at the University of Colorado at Boulder. Instead, on the advice of his girlfriend, to whom I am forever grateful, he went to art school in Baltimore, studied photography, then set himself up in Brooklyn, NY, to do carpentry, mostly building kitchens for New York homes. On the side he designed some furniture and had a very fruitful year as a designer with a large furniture producer in Pennsylvania. In 1975, he designed the Spector NS model electronic bass guitar, the same design still sold today. The design of electronic musical instruments has been his work since. Two years later, he designed the "headless" graphite Steinberger Bass, a very original, novel design, which won him a great deal of recognition in electronic guitar circles. It was recognized in 1980 by the Industrial Designers Society of America Excellence Award; in 1981, it received the Time Magazine award as one of the ten best designs of the year; and in 1989, the Industrial Designers Society of America Design of the Decade Award.

In 1979, with the financial support of several friends, he started the Steinberger Sound Corp. to produce the bass and guitars designed along the same lines, the bodies built with the very rigid material carbon fiber. He directed this enterprise in Newburgh-on-the-Hudson, with perhaps 30 employees, and managed to survive until Gibson bought the company some ten years later. The Steinberger instruments were almost handmade and quite expensive; in the meantime, Ned licensed the design to other companies in Japan and Germany, which made similar instruments but of more conventional materials. The sale of Steinberger Sound for the first time gave Ned some financial independence, and he was able to design for himself and build a house that I admired a great deal for its originality, simplicity, and practicality. At 45, Ned found and married Denise, a medical doctor, in Damariscotta, Maine.

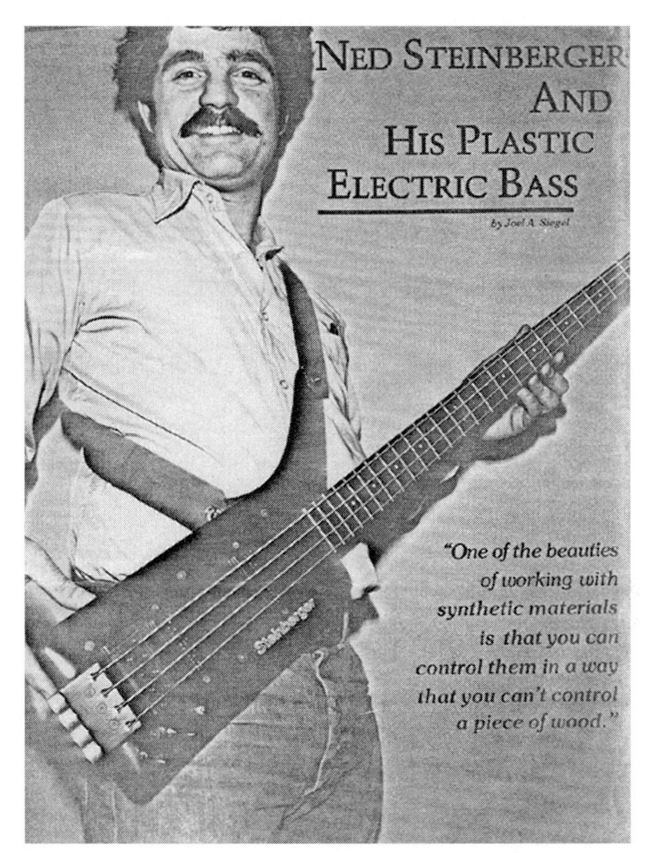

Fig. A.1. Ned, with one of his guitars. Picture used in a trade magazine (Credited to Alan Gaul, Guitar World)

They have two children, and he designs and collaborates in the construction of beautiful electronic string basses, cellos, and violins.

Hobbies

Tennis. I managed to keep reasonably fit most of my life. Tennis I enjoyed quite a bit until arthritis stopped this at the age of about 60. It began when, after high school graduation, I moved in with my parents in Rogers Park, in the north of Chicago, and, to earn a few pennies, rolled some tennis courts on the other side of the "L" (Chicago's elevated subway) from our Jewish delicatessen store. The owners of the courts were kind enough to introduce me to the game.

Mountaineering. More exiting was mountaineering. My first contact with mountains came in 1956 when I attended a summer school organized at Les

Houches, at the foot of Mont Blanc, by my friend from my days at the Princeton Institute of Advanced Studies, Cecile Morette. Her husband, also a physicist, specializing in relativity, took a few of us first to the Aiguille du Midi with the telepherique, then down and around to the Col du Midi, where there is a hut, constructed before the war as a cosmic ray laboratory but now a mountaineering hut, and then down the Mer-de-Glace. The summer of 1968, as visitor at the University of Washington in Seattle, WA, I had the luck to find a group of graduate students, active mountaineers, who gave me my first lessons in the art and took me on some fine climbs in the Cascades: Mt. Rainier, Mt. Hood, Mt. Baker, and, the most fun of all, Mt. Shukson. By chance, one or two years later, Franz Mohling, one of the most capable of the Seattle mountaineering group, came to our physics department at Columbia and sometimes took me to practice rock climbing on the Shawangunk cliffs, some hundred miles north of New York City. Several summer visits to Geneva, the 1.5 year sabbatical, '65–'66, and the final move in '68 opened the Alps, and especially spring ski touring with skins. In Varenna, on Lake Como, where the Italian Physical Society hosted summer schools, I was able to make some friends, members of a nearby Club Alpino Italiano, who took me climbing on the nearby Grigna cliffs and on a memorable tour to the Piz Bernina. I was very addicted, but not particularly competent, either in rocks or skiing.

Mountaineering has its risks and three of my close friends, including Franz, became its victims. I had three difficult calls myself. The first, coming down from the Mont Blanc on a ski tour, I fell on an icy, reasonably steep glacier, and on sliding down, took with me my companion Paul Actis, for maybe 75 meters. We flew over a substantial crevasse before coming to a stop, on its downhill side, by chance safe and sound. Some years later, Paul fell victim to the mountains on a climb in the Aiguilles Rouges, near Chamonix. Another memorable hour was on the Grepon, a well known, beautiful climb above Chamonix, close to Geneva. I was with Alfred Thissières, a biology professor at the University of Geneva who was known for some first ascents in his younger years in the Valais Alps, and at the time of our climb he was no longer so young. I got stuck in a famous chimney, the Walker Spur, was perfectly exhausted, and we could not move until perhaps half an hour later when a passing team helped us get out of this. Finally, I managed to fall into a crevasse. With a younger friend we were on our way to the Argentière hut, above the Argentière glacier, to climb the Argentière the next day. We were in shorts, had just crossed the very flat glacier, and were on the moraine, just below the hut, when we heard a distress whistle. We went back to the glacier, following the sound. As we moved on, the condition of the glacier became a bit more worrying and, luckily for me, we stopped to put on the rope. Continuing, on jumping a crevasse, I slid back in, descending between two slippery vertical walls, my back against one side, the shoes on the other, maybe ten meters, maybe 15, until I was stopped by the rope and a fortunate slight outcropping on the face side of the crevasse. My companion had been sliding on the horizontal ice above. There we were. I didn't look down to see

Fig. A.2. Climbing practice on the Shawangunk cliffs, north of New York, late fifties (with permission by Jack Steinberger)

how deep it was. We were saved by the guardian of the hut and his son, who had also heard the whistle and who pulled me out before moving on to help the other crevasse victim. A helicopter later came to take the latter, with fractured skull, to the hospital. I escaped with minor scratches. Ted did the Argentière the next day, but I stayed in the hut. A year or two later, I returned with other friends to climb this mountain, but I had become much warier, and stopped climbing altogether in about 1975.

Why do people climb? During the sixties at CERN, I knew an electronic technician, a very good mountaineer, who when climbing was evidently thrilled by the beauty of the surroundings. At a certain moment he quit his job to become a professional mountain guide. But, in general, the people I have observed in the mountains are more motivated by the challenge than by the beauty. They are more interested in showing to themselves, and perhaps to others, that they can do something that is difficult, which not everybody can do. There is an interesting parallel with scientific research. We scientists should be motivated by the desire to understand nature a bit better, but often, what really drives us is the need to show to ourselves and to our colleagues that we can do this better than the other, to be the first in publishing a result.

Fig. A.3. With Hallstein Hogasen on the Aiguille du Plan, near Mont Blanc, ~ 1970

Sailing. When I was 20 and had to drop college (for a year) to earn a living, my job and my work in the family store evenings and weekends did not keep me from building a ten-foot sailing dinghy out of plywood. It had been designed by an acquaintance who was building his own. A few years later, during the war, at MIT, I could participate in the sailing and racing on the Charles River with the famous MIT dinghies from the school boathouse. I managed, by some stupidity, to seriously damage one of these. This was the end of sailing until, in 1975, by some accident, I bought JuliaK., an Etap 22 ft (6.6 m) sloop, unsinkable (it was stuffed with enough plastic foam), a retractable keel, sleeping accommodations for three, a potty, an alcohol burner, and a 6 HP outboard motor. Julia was a year old, so the sailing was with friends, not family, along the Mediterranean coast: Corsica, the Balear Islands, for a month each summer. In the winter, JuliaK. was on a trailer in front of our house at the foot of the Jura mountains, overlooking Mont Blanc. The Balear trip was my first visit to Spain; I could not get myself to go there as long as Franco was in power. Once Cynthia and two-year-old Julia came along for a few days along Corsica.

In 1979, we were four, John was 2, and JuliaK. was replaced by the larger Zig Zag, Julia's choice of name, a 31-foot Hallberg Rassy Monsun, very seaworthy, with adequate accommodations for six, a real toilet, a pantry, a gas stove, and an 18hp diesel engine, but no longer trailerable. The family, includ-

Fig. A.4. Retrieving Zig Zag at her winter quarters, Ampuria Brava, Spain, summer 1981

ing John, came along to fetch Zig Zag at the boatyard, on Ellös Island, on Sweden's west coast. With the help of friends, the Tibells from Uppsala, we sailed her down the beautiful Swedish Schären-coast to Anholt, across to Jutland, and through the Kiel Kanal, down the Elbe River to Cuxhaven. There, the family was replaced by friends, and Zig Zag continued down the Friesien coast, after having been blown to Helgoland by head winds too strong for us to manage, then through some beautiful Dutch canals, through the Ijselmeer and Amsterdam, and across the British Channel to Southhampton. The next year, Zig Zag went around the Spanish and Portuguese coasts, after being evicted from Gibraltar by an officer of Her Majesty's navy, to winter in a harbor in Catalonia, close to France. The following year, it was to Italy, around Corsica, and then, finally back to Toulon.

Zig Zag was replaced by the bigger Sadler 34-foot Zig Zag II. I cannot understand now, nor remember, what motivated this change; the Monsun was an excellent cruising boat and adequate in the matter of accommodations. As it turned out, I regretted the change; Zig Zag II was too big for me to handle in ports, and I managed to damage another boat with her. Zig Zag II was boarded in Poole and sailed to the Baltic, where she stayed, wintered on Fehmarn island, until '88, when she was sold because the next summer I wanted to be free to participate in the birth pains of our detector Aleph, by that time eight years in design and building, at the start-up of the electron–positron collider LEP at Cern. I had appreciated very much the sailing around the Danish islands, with many peaceful harbors in old villages. In particular, Copenhagen, not really a village, has a beautiful old, historic harbor, and it is possible for visiting boats to make up just outside of this.

The year after the start-up of LEP, life was again more relaxed. A new, final boat, Zig Zag III, was bought, again a Hallberg Rassy 31, but now a different design, and a new, four-summer trip began, from Ellös, through the Baltic and Dutch canals, past England, around Spain, and to Italy. One of the most enjoyable moments was the visit to Lisbon, a beautiful, historic city with ancient ports. The trip up the Tagus estuary, under the famous, splendid hanging bridge, and past the Belem castle and the monument to Prince Henry the Navigator, built by Salazar the dictator, was memorable. The summer of 1994 marked the end of sailing days, but it has been a special thrill for me to get Christmas cards with stamps from exotic islands in the Pacific from Zig Zag III's new owners, Manfred Krischak and Rita Martin, who are sailing her around the world.

Music. Father was a competent pianist who played the classics. On the side, he gave piano lessons, but for reasons not known to me, he taught none of his children. At 23, during the war, while working at the MIT radar development laboratory, I started taking piano lessons from a private teacher, a lady, fellow refugee from the Nazis, and also a course in music at the New England Conservatory. After the war and graduate studies, once settled down at Columbia University, I continued and tried to play Haydn and easier Mozart and Beethoven sonatas, but with remarkable, frustrating lack of success. A turn for the better in my musical career I owe to the ill-famed Senator McCarthy, a story I told in Chap. 4, when the conductor, Max Goberman, victim of McCarthy, advised me to change to the recorder.

This advice changed my life. For one, it was easier for me to make some headway with the recorder, which a year or two later was replaced by the transverse flute, but chiefly because it made it possible to get together with others to play chamber music. Doing music with others is not only an intellectual, it is also a social pleasure. I acquired a harpsichord. Being near New York, and with some luck, I found friends to play chamber music, which included occasionally also an evening to do arias with obligato accompaniment, from Bach cantatas. This has continued to the present. Eventually, I could also play a bit of the classics, and the harpsichord was replaced by a grand piano. Sometimes we would be perhaps eight musicians and could do Bach Brandenburg Concerti. The greatest moment came in 1988. When our Geneva "commune," Onex, learned of the Nobel Prize of one of its residents, it organized a reception. There I met the mayor of Onex, Jean-Claude Cristin, who was also first French horn in the Orchestre de la Suisse Romande. Among my musical friends there was a wonderful American-Belgian, Oboe–French horn professional musician couple. The mayor agreed to come, and for the first and only time we could do the 2nd Brandenburg, with two French horns, three oboes (my flute was one of these), strings, and piano. The sound of the solo trio of the two horns in our living room was, and is, one of my more memorable pleasures.

Music is important for me, especially now, in old age. I enjoy renaissance music, baroque, the classics, especially chamber music, the romantics, espe-

Fig. A.5. Chamber music at home in Gex

cially Shubert, and into the early 20th century, for instance, Mahler. My favorite is Bach, and since some years, on Sundays, I offer myself an hour or two of his cantatas, on discs. If I had to choose the all-time greatest, most creative human being, it would be Bach.

Opinions

For what it may be worth, I would like to add here my opinion on some contemporary social issues.

Science and Society. Scientific progress changes our society. Some of the changes may be positive, for instance, recent progress in medicine, informatics, communication and travel, energy production and agriculture. Some of the progress is more questionable such as the invention of nuclear weapons or of human cloning. But even the progress generally considered positive may raise difficult problems for society. Medical progress raises the problem of spiraling medical costs and the burden on society of an aging population; travel and energy production raise the ecological problems of consumption of irreplaceable resources, global warming or nuclear waste hazards; the overall improvement in human living conditions raises the important and difficult problem of population explosion, etc. What should be the attitude of the scientist, how should he direct his work so that humanity might profit rather than suffer? My answer to this is the following, and here I need to differentiate between basic and applied science: Basic science, the effort to understand our universe, is a cultural enterprise. To the extent that we value knowledge, any progress in our basic understanding is good. The researcher is not able, generally, to foresee the possible applications of his progress and, therefore, cannot be held responsible for its future uses, if any, good or bad. The responsibility for the possible applications is with society as a whole. Technical progress resulting from new basic insights may pose important, difficult questions to society and governments. Even applied research may pose this problem, since the impact of the application may also not be adequately foreseeable. But sometimes the scientist does have a choice. I would rather work on the design of renewable energy sources, or on the storage or treatment of nuclear waste, than on the design of new nuclear weapons.

Nuclear Disarmament. Nuclear weapons are a serious threat to the future of mankind. Biological weapons are also a substantial, less well understood danger, but, as best I know, nuclear weapons are by far the more dangerous. The present arsenals are capable of destroying our civilization. Some impression of the destructive power may be possible from the figure which compares the world's nuclear explosive power, some 20,000 megatons, with the total of three megatons used in World War II. The threat of cataclysm is real. Recently, on the occasion of the 40th anniversary of the Cuban Missile Crisis of 1962, we were told by the then U.S. Secretary of Defense Robert S. McNamara that we came very close to the actual nuclear showdown, much closer than we were told at the time.

Nuclear weapons threaten civilization, but serve no realistic purpose. Even though the U.S. government, in recent policy statements, has proposed that its nuclear weapons might be used against a threat of other, non-nuclear weapons of mass destruction, or even preemptively, it is difficult to imagine that this would be a reasonable policy. Each country has more to lose in the use of nuclear weapons than it can gain, most of all the U.S., invulnerable to anything but nuclear weapons.

Nuclear weapons could be eliminated. If, instead of its present nuclear weapons policy, the U.S. would decide to move forthrightly toward the global elimination of nuclear weapons, the other nuclear-weapon states would follow. This is the stated policy of China: India says that it has its nuclear weapons to deter China and so would follow China, and so Pakistan with India. Russia, under the pressure of its difficult economic situation, has clearly shown interest in nuclear disarmament. Even Israel could not resist a U.S. move toward global nuclear disarmament.

Nuclear disarmament could be verified. The construction of nuclear weapons requires large, specialized facilities that could be observed. The greatest verification problem, to my knowledge, would be possible dissimulation of nuclear warheads by present nuclear weapons states. But any such move would risk the disclosure by one of the many persons necessarily involved and informed of such an act.

In the meantime, the problem of proliferation marches on. In the Nonproliferation Treaty of 1970 the nuclear powers committed themselves to nuclear disarmament. When the treaty came up for renewal in 1995, the nuclear

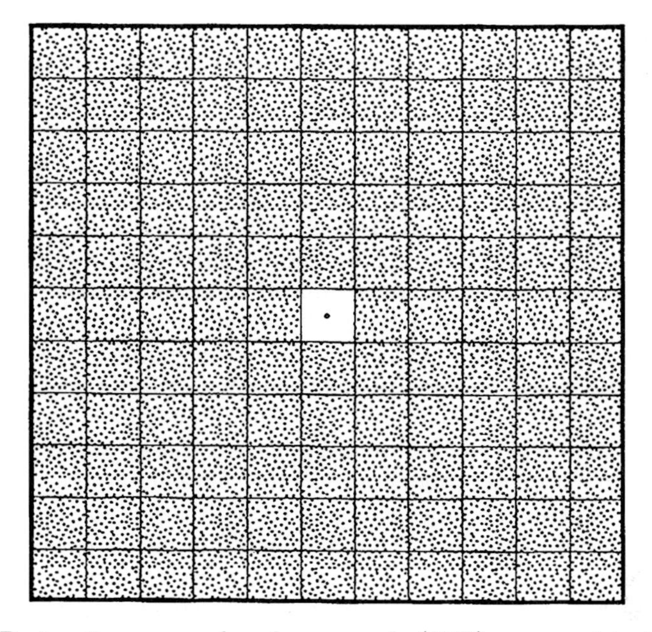

Fig. A.6. Destructive power of nuclear arsenals (1983) compared to that used in World War II, represented by the dot in the center, three megatons of TNT. The 18,000 megatons TNT equivalent of 1983 arsenals correspond to 6,000 of World War II's (Science and Society, 12 (1983) 12)

weapons stockpiles instead had increased. India made its signature for treaty renewal conditional on agreement by the nuclear weapons states to a timetable for disarmament. The United States declined, now India and Pakistan have nuclear weapons. Will North Korea be next? The problem of the nuclear threat will increase with time, until the nuclear powers move toward total nuclear disarmament. This is the only possible resolution of this terrible danger, and America must lead the way. It is the country which first produced them, the country which has used them, by far the greatest power in the world, and the country which has most to gain by their elimination.

Nuclear and other energy. The world's population has tripled within the last century, and per capita energy consumption has also roughly tripled. Present world energy consumption is dominated largely by fossil fuel use. Within a few generations we are exhausting that part of the global fuel resources which is easily accessible. The CO_2 production due to this use is changing the earth's climate substantially, which threatens, in particular, the populations of the warmer regions of the globe, in the near future, and for many hundreds of years thereafter. This is a double crime, by our generation against future generations.

There are those who propose "sustainable development." Presumably "development" means yearly increase in economic production. It seems that in order for our economies to survive, to be competitive, production must increase several percent each year. This by definition is an exponential growth, and no exponential growth is sustainable, so I consider the proposition of "sustainable development" to be a contradiction in terms, a sham. We are simply using much too much energy, at the expense of our children's children. There are no alternatives, no non-polluting energy sources at the present level of energy consumption. The only logical solution is a drastic reduction in energy consumption, particularly in the industrialized world. But our society, in particular, our economy, is not organized in a way to make this a realistic possibility. Clearly, I am pessimistic about the future, starting perhaps 50 years from now.

Could nuclear energy help? Nuclear fusion energy is not yet on the horizon. Fission now produces only one or two percent of the global energy consumption, so the replacement of fossil fuels by nuclear energy would require a huge increase in nuclear energy production; furthermore, its use for transportation is technically difficult, not obvious. In addition, there is the problem of long-lived nuclear waste, as well as the vulnerability of nuclear plants as terrorist targets. Nevertheless, I would prefer to see the use of nuclear energy to that of fossil fuels, the likely damage to future generations would be less by a large factor. Unfortunately, I expect that we will first exhaust cheap fossil fuels, with consequent global warming, and then turn to the nuclear power.

Politics, economics. I have always been sympathetic to the underprivileged, to the Left, however have never been a member of any party. During the war, as a very young man, I joined a labor union, the Union of Scientific Workers, CIO, and campaigned for the reelection of Roosevelt. My leftist sympathies, during the Cold War and McCarthy years, earned me some persecution by the FBI, which I have already related. I am very concerned about the inequalities inherent in our capitalist society, between the owners and the workers, the haves and the have nots, ever increasing. Even if capitalist competitiveness were the most efficient, I would still prefer lower productive efficiency and more equality. Perhaps the best system I have seen was that of Sweden in the '60s and '70s. But in our global economy, it is very difficult for a country to maintain a high level of socialism. On the other hand, the "dictatorship of the proletariat" has not succeeded either. In China, it brought some good things, such as a high level of equality (now in the process of being destroyed), as well the emancipation and education of women, but it also brought the horrors of the "cultural revolution." I am also of the opinion that our competitive capitalism is incompatible with an ecologically sustainable economy and therefore globally very destructive. However, I must admit that, now in my old age, I am at a loss to propose the politico-economic global structure that would adequately serve and protect our society. I consider the solution of this problem the most important challenge to our society.

References

1. H.-J. Beck and R. Walter, Jüdisches Leben in Bad Kissingen.
2. C. N. Yang, Selected papers 1945–1980 with commentary, W. H. Freeman and Company, San Francisco, 1983.
3. M. Conversi, E. Pancini, and O. Piccioni, Phys. Rev. **71** (1947) 209.
4. C. G. M. Lattes, H. Muirhead, G. P. S. Occhialini, and C. F. Powell, Nature **159** (1947) 694.
5. G. D. Rochester and C. C. Butler, Nature **160** (1947) 885.
6. J. Steinberger, On the range of electrons in muon decay, Phys. Rev. **75** (1949) 1136.
7. J. Tiomno and J. A. Wheeler, Rev. Mod. Phys. **21** (1949) 144.
8. T. D. Lee, M. Rosenbluth, and C. N. Yang, Phys. Rev. **75** (1949) 905.
9. G. Puppi, Nuovo Cimento **VI** (1949) 194.
10. B. Pontecorvo, Phys. Rev. **72** (1947) 246.
11. W. Pauli and F. Villars, Rev. Mod. Phys. **21** (1949) 433.
12. J. Steinberger, Phys. Rev. **76** (1949) 1180.
13. E. Fermi and C. N. Yang, Phys. Rev. **76** (1949) 1739.
14. S.L. Adler, Phys. Rev. **177** (1969) 2426, and J. S. Bell and R. W. Jackiv, R. Nuovo Cimento **604** (1967) 47.
15. P. Bjorklund, W. E. Crandall, B. J. Moyer, and H. York, Phys. Rev. **77** (1950) 213.
16. L. W. Alvarez et al., Proc. Am. Phys. Soc., Phys. Rev. **77** (1950) 752.
17. J. Steinberger and A. S. Bishop, Phys. Rev. **78** (1950) 494.
18. J. Steinberger, W. K. H. Panofsky, and J. Steller, Phys. Rev. **78** (1950) 802.
19. O. Chamberlain, R.F. Mozeley, J. Steinberger, and C. Wiegand, Phys. Rev. **79** (1950) 394.
20. C. Chedester, P. Isaacs, A. Sachs, and J. Steinberger, Total cross-sections of π^- mesons on protons and several other nuclei, Phys. Rev. **82** (1951) 958.
21. R. E. Marshak, Phys. Rev. **82** (1951) 313.
22. Cartwright, Richman, Whitehead, and Wilcox, The production of π^+ mesons by protons on protons, Phys. Rev. **81** (1951) 652, and V. Peterson, Phys. Rev. **79** (1950) 407.
23. R. Durbin, H. Loar, and J. Steinberger, The spin of the pion via the reaction $\pi^+ + d \rightleftharpoons p + p$, Phys. Rev. **83** (1951) 646. See also Clark, Roberts, and Wilson, Phys. Rev. **83** (1951) 649.

24. P. Lindenfeld, A. Sachs, and J. Steinberger, Phys. Rev. **89** (1952) 531.
25. W.K.H. Panofsky, R.L. Aamodt, and J. Hadley, The gamma ray spectrum resulting from capture of negative π-mesons in hydrogen and deuterium, Phys. Rev. **81** (1951) 565.
26. R.H. Dalitz, Proc. Phys. Soc. (London) **A64** (1951) 667.
27. W. Chinowsky and J. Steinberger, Phys. Rev. **93** (1993) 586.
28. H.L. Anderson, E. Fermi, R.D. Martin, and D.E. Nagle, Phys. Rev. **91** (1953) 155.
29. D. Bodansky, A.M. Sachs, and J. Steinberger, Phys. Rev. **93** (1954) 1367.
30. J. Ashkin, J. Blaser, F. Feiner, and M.O. Stern, Phys. Rev. **101** (1956) 1149.
31. J. Ashkin, J. Blaser, F. Feiner, J.G. Gorman, and M.O. Stern, Phys. Rev. **93** (1954) 1129.
32. W. Chinowsky and J. Steinberger, Phys. Rev. **95** (1954) 1561.
33. N. Kroll and W. Wada, Phys. Rev. **98** (1955) 1355.
34. Plano, Prodell, Samios, Schwartz, and Steinberger, Phys. Rev. Lett. **3** (1959) 524.
35. S. Lokanathan and J. Steinberger, Nuovo Cimento **10** (1955) 151.
36. Fazzini, Fidecaro, Merrison, Paul, and Tollesstrup, Phys. Rev. Lett. **1** (1958) 247.
37. Impeduglia, Plano, Prodell, Samios, Schwartz, and Steinberger, Phys. Rev. Lett. **1** (1958) 249.
38. R.W. Thompson, Proceedings of the 3rd Rochester Conference **39** (1952), and Thompson, Buskirk, Etter, Katzmark, and Redicker, Phys. Rev. **90** (1953) 329.
39. A. Pais, Some remarks on the V particles, Phys. Rev. **86** (1952) 663.
40. M. Gell-Mann, Isostopic spin and new unstable particles, Phys. Rev. **92** (1953) 833.
41. W.B. Fowler, R.P. Shutt, A.M. Thorndike, and W.L. Whittemore, Production of heavy particles by negative pions, Phys. Rev. **93** (1956) 861.
42. M. Gell-Mann and A. Pais, Behaviour of neutral particles under charge conjugation, Phys. Rev. **97** (1955) 1387.
43. K. Lande, E.T. Booth, J. Impeduglia, L.M. Lederman, and W. Chionowsky, Observation of long-lived neutral V-particles, Phys. Rev. **103** (1956) 1901.
44. D.A. Glaser, Bubble chamber tracks of cosmic ray particles, Phys. Rev. **91** (1953) 762.
45. J.G. Wood, Phys. Rev. **94** (1954) 731.
46. J. Leitner, N. Samios, M. Schwartz, and J. Steinberger, Nevis Report 10, 1955.
47. D.A. Glaser and D.C. Rahm, Phys. Rev. **97** (1955) 474.
48. Budde, Chretien, Leitner, Samios, Schwartz, and Steinberger, Properties of heavy unstable particles produced by 1.3 BeV π^- mesons, Phys. Rev. **103** (1956) 1827.
49. T.D. Lee and C.N. Yang, Phys. Rev. **104** (1956) 254.
50. Wu, Ambler, Hayward, Hoppes, and Hudson, Phys. Rev. **105** (1957) 1423.
51. O. Chamberlain, E. Segre, C. Wiegand, and T. Ypsilantis, Observations of antiprotons, Phys. Rev. **100** (1955) 947.
52. V.E. Barnes et al., Observation of a hyperon with strangeness minus three, Phys. Rev. **12** (1964) 204.
53. R. Plano, N. Samios, M. Schwartz, and J. Steinberger, Demonstration of the existence of the $\sum^\circ$ hyperon and a measurement of its mass, Nuovo Cimento V, (1957) 216.

54. Eisler, Plano, Samios, Schwartz, and Steinberger, Systematics of Λ° and θ° decay, Nuovo Cimento V, (1957) 1700.
55. F. Eisler et al., Demonstration of parity non-conservation in hyperon decay, Phys. Rev. **108** (1957) 1353.
56. F. S. Crawford et al., Detection of parity non-conservation in Λ decay, Phys. Rev. **108** (1957) 1102.
57. F. Eisler et al., Experimental determination of the Λ° and $\sum^-$ spins, Nuovo Cimento **VII**, (1958) 222.
58. R. Adair, Phys. Rev. **100** (1955) 1540.
59. A. R. Erwin, R. March, W. D. Walker, and E. West, Evidence for a $\pi - \pi$ resonance in the I=1, J=1 state, Phys. Rev. Lett. **6** (1961) 638.
60. Allston, Alvarez, Eberhard, Good, Graziano, Ticho, and Wojcicki, Resonance in the Lp system, Phys. Rev. Lett. **5** (1960) 520.
61. N. Gelfand et al., Lifetime of the ω meson, Phys. Rev. Lett. **11** (1963) 436.
62. N. Gelfand et al., Width of the ϕ meson, Phys. Rev. Lett. **11** (1963) 438.
63. C. Alff et al., $\sum^\circ - \Lambda^\circ$ relative parity, Siena 1963 Conference Report, p. 205.
64. H. Courant et al., Phys. Lett. **10** (1963) 409.
65. J. Chadwick, Verh. d. deutschen Phys. Ges. **16** (1914) 383.
66. C. D. Ellis and W. A. Wooster, Proc. Roy. Soc. (a) **117** (1927) 109.
67. J. Chadwick, Nature **129** (1932) 312.
68. E. Fermi, Zeitschrift für Physik **88** (1934) 161.
69. R. D. Albert and C. S. Wu, Phys. Rev. **74** (1948) 847.
70. F. Reines and C. L. Cowan, Nature **178** (1956) 446.
71. G. Feinberg, Phys. Rev. **110** (1958) 1482.
72. B. Pontecorvo, J. Exptl. Theoret. Phys. (U.S.S.R.) **37** (1959) 17561.
73. M. Schwartz, Phys. Rev. Lett. **4** (1960) 306.
74. T. D. Lee and C. N. Yang, Phys. Rev. Lett. **4** (1960) 307.
75. Danby, Gaillard, Goulianos, Lederman, Mistry, Schwartz, and Steinberger, Observation of high energy neutrino reactions and the existence of two kinds of neutrinos, Phys. Rev. Lett. **9** (1962) 36.
76. F. J. Hasert et al., Observation of neutrino-like interactions without muon or electron in the Gargamelle neutrino experiment, Phys. Lett. **45B** (1973) 138.
77. G. Arnison et al., Experimental observation of isolated large transverse energy electrons with associated missing energy at $\sqrt{s} = 540$ GeV, Phys. Lett. **122B** (1983) 103.
78. G. Arnison et al., Experimental observation of lepton pairs of invariant mass around 95 GeV, Phys. Lett. **126B** (1983) 398.
79. J. H. Christenson, J. W. Cronin, V. L. Fitch, and R. Turlay, Evidence for the 2π decay of the K_2^0 meson, Phys. Rev. Lett. **13** (1964) 138.
80. T. T. Wu and C. N. Yang, Phys. Rev. Lett. **13** (1964) 501.
81. C. Alff-Steinberger et al., K_S and K_L interference in the $\pi^+\pi^-$ decay mode, CP invariance and the $K_S - K_L$ mass difference, Phys. Lett. **20** (1966) 207 and Phys. Lett. **21** (1966) 595.
82. V. L. Fitch, R. F. Roth, J. S. Russ, and W. Vernon, Evidence for constructive interference between coherently regenerated and CP-nonconserving amplitudes, Phys. Rev. Lett. **15** (1965) 73.
83. Bennett, Nygren, Saal, Steinberger, and Sutherland, Measurement of the charge asymmetry in the decay $K_L \rightarrow \pi^{+/-} + e^{-/+} + \nu$, Phys. Rev. Lett. **19** (1967) 993.

84. Dorfan, Enstrom, Raymond, Schwartz, Wojcicki, Miller, and Paciotti, Charge asymmetry in the muonic decay of the K_L, Phys. Rev. Lett. **19** (1967) 987.
85. Bennett, Nygren, Saal, Sutherland, and Steinberger, Measurement of the $K_S - K_L$ regeration phase in copper at 2.5 GeV/c, Phys. Lett. **27B** (1968) 239.
86. J. S. Bell and J. Steinberger, Weak interaction of kaons, in Proceedings of the Int. Conf. on Elementary Particles (Rutherford Laboratory, Chilton, England, 1965).
87. M. Banner, J. W. Cronin, J. K. Liu, and J. E. Pilcher, Measurement of the branching ratio $(KL \to 2\pi^\circ)/\ (KL \to 3\pi^\circ)$, Phys. Rev. Lett. **21** (1968) 1107.
88. J. C. Chollet et al., Observation of the interference between K_L and K_S in the $\pi^\circ\pi^\circ$ decay mode, Phys. Lett. **31B** (1970) 658.
89. Gibbons et al., Phys. Rev. Lett. **70** (1993) 1199, Schwingenheuer et al., Phys. Rev. Lett. **74** (1994) 376.
90. G. Charpak, R. Bouclier, T. Bresani, J. Favier, and C. Zupancic, Nucl. Instr. and Meth. **62** (1968) 262.
91. A. H. Wallenta, J. Heintze, and B. Schnerlein, The multiwire drift chamber, Nucl. Instr. and Meth. **92** (1971) 373.
92. P. Schilly et al., Construction and performance of large multiwire proportional chambers, Nucl. Instr. and Meth. **91** (1970) 221.
93. W. C. Carithers et al., Observation of the decay $K_L \to \mu^+ + \mu^-$, Phys. Rev. Lett. **30** (1973) 1336.
94. C. Geweniger et al., Measurement of the charge asymmetries in the decays $K_L \to \pi + e + \nu$, and $K_L \to \pi + \mu + \nu$, Phys. Lett. **48B** (1974) 483.
95. S. Gjesdal et al., A measurement of the $K_S - K_L$ mass difference from the charge asymmetry in semileptonic kaon decays, Phys. Lett. **52B** (1974) 113.
96. C. Geweniger et al., A new determination of the $K^0 \to \pi^+\pi^-$ parameters, Phys. Lett. **48B** (1974) 487.
97. C. Geweniger et al., Measurement of the kaon mass difference by the two regenerator method, Phys. Lett. **52B** (1974) 108.
98. S. Gjesdal et al., The phase of φ_{+-} of CP violation in the $K^0 \to \pi^+\pi^-$ decay, Phys. Lett. **52B** (1974) 119.
99. S. Gjesdal et al., A measurement of the total cross-section of lambda hyperons on protons and neutrons in the momentum range 6 GeV/c to 21 GeV/c, Phys. Lett. **40B** (1972) 152.
100. F. Dydak et al., Measurement of the $\sum^\circ$ lifetime, Nucl. Phys. **B118** (1977) 1.
101. F. Dydak et al., Measurement of the electromagnetic interaction of the neutral kaon, Nucl. Phys. **B102** (1976) 253.
102. A. Apostolakis et al., CPLear, SA determination of the CP violation parameter η_{+-} from the decay of strangeness tagged neutral kaons, Phys. Lett. **B458** (1999) 545.
103. H. Burkhardt et al., First evidence for direct CP violation, Phys. Lett. **B206** (1988) 169.
104. G. D. Barr et al., A new measurement of direct CP violation in the neutral kaon system, Phys. Lett. **B317** (1994) 233.
105. L. K. Gibbons et al., Measurement of the CP violating parameter $\mathrm{Re}(\epsilon'/\epsilon)$, Phys. Rev. Lett. **70** (1993) 1203.
106. J. R. Bathley et al. (CERN NA48 Collaboration), A precision measurement of direct CP violation in K decay, Phys. Lett. **B544** (2002) 97.
107. A. Halavi-Harati et al. (Fermilab KTeV Collaboration), Measurement of direct CP violation and CPT symmetry in K° decay, Phys. Rev. **D67** (2003) 012005.

108. K. Abe et al. (Belle Collaboration), Improved measurement of mixing induced CP violation in the neutral B meson system, Phys. Rev. **D66** (2002) 071102.
109. B. Aubert et al. (BaBar Collaboration), Measurement of the CP asymmetry amplitude sin2β with B^0 mesons, Phys. Rev. Lett. **89** (2002) 201802.
110. H. Breidenbach et al. (Stanford Linear Accelerator), Observed behaviour of highly inelastic electron-proton scattering, Phys. Rev. Lett. **23** (1969) 935.
111. H. Deden et al., Gargamelle, Structure functions in charge changing interactions of neutrinos and anti-neutrinos, Nuclear Physics **B85** (1975) 269.
112. M. Holder et al., Is there a high y anomaly? Phys. Rev. Lett. **39** (1976) 433.
113. M. Holder et al., Study of inclusive neutral current interactions of neutrinos and antineutrinos, Phys. Lett. **72B** (1977) 254.
114. H. Abramowitz et al., Neutrino and anti-neutrino inclusive scattering in iron in the energy range 20–300 GeV, Particles and Fields **17** (1982) 283.
115. M. Holder et al., Opposite sign dimuon events produced in narrow band neutrino and antineutrino beams, Phys. Lett. **70B** (1977) 393.
116. M. Holder et al., Inclusive interactions of neutrinos and antineutrinos in iron. Particle and Fields **1** (1979) 143.
117. H. Abramowitz et al., Determination of the gluon distribution in the nucleon from deep inelastic neutrino scattering, Particles and Fields **12** (1982) 289.
118. B. Pontecorvo, Neutrino experiments and the problem of conserving leptonic charge, Zh. Eksp. Teor. Fiz. 53, (1967) 1717, and V. Gribov and B. Pontecorvo, Neutrino atronomy and leptonic charge, Phys. Lett. **28B** (1968) 493.
119. F. Dydak et al., Search for muon neutrino oscillations, Phys. Lett. **134B** (1983) 281.
120. S. Fukunda et al., Super Kamiokande Coll., Phys. Rev. Lett. **86** (2001) 5651.
121. S. Fukunda et al., Super Kamiokande Coll., Phys. Rev. Lett. **B476** (1999) 185 and Phys. Rev. Lett. **82** (1999) 2644.
122. C. B. Netterfield et al., Measurement by Boomerang of multiple peaks in the CMBR power spectrum, ApJ **517** (2002) 604.

Printing: Strauss GmbH, Mörlenbach
Binding: Schäffer, Grünstadt